PORTLAND PRESS RESEARCH MONOGRAPH

# The Annexins

11

PORTLAND PRESS RESEARCH MONOGRAPH II

# The Annexins

Edited by
**Stephen E. Moss**

**Portland Press**
**London and Chapel Hill**

Published by Portland Press Ltd, 59 Portland Place, London W1N 3AJ, U.K.
In North America orders should be sent to Portland Press Inc.,
P.O. Box 2191, Chapel Hill, NC 27515–2191, U.S.A.



**ISBN 1 855 78 008 9 ISSN 0964–5845**

**British Library Cataloguing in Publication Data**
A catalogue record for this book is available from the British Library



Printed in Great Britain by the University Press, Cambridge

*Front cover*: Crystal structure of annexin V
(R. Huber, Max-Planck Institute, Martinsried)

# Preface

$Ca^{2+}$-binding proteins are of great intrinsic interest because of their prospective biological importance. When such proteins are expressed within both plant and animal kingdoms, including invertebrates and vertebrates, and have been highly conserved during evolution, their anticipated biological importance, and thus their associated interest, becomes considerable. The above set of circumstanes applies to the annexin supergene family of $Ca^{2+}$- and phospholipid-binding proteins, various members of which have been described by several different laboratories over the past decade. The motivations underlying their discovery have varied within the different laboratories and this has in turn been reflected in the names ascribed to both individual members and the family. They have been called variously lipocortins, calpactins, calelectrins, calcimedins, chromobindins, endonexins and vascular anticoagulant proteins. This babylonic mixture of names has inevitably led to confusion and, through fragmentation, has inhibited growth of the field. A common nomenclature based on amino acid sequence homology has led to the adoption by most investigators of the generic term 'annexin'.

The annexin family presently consists of some 13 members expressed in organisms as widely diverse as higher plants, slime moulds, metazoans, insects, birds and mammals. Each member is characterized by having a conserved 70-amino-acid repeating unit which is repeated four times, with the exception of annexin VI which has eight repeats. It is anticipated that the family arose from a primodial gene coding for a single repeat that underwent duplication and diversification during evolution. If this view is correct, then proteins containing a single or two repeats would be surmised to exist, but such proteins have yet to be discovered. The main sequence variations are located at the *N*-terminus which also shows considerable variation in length. Given the above overall structural framework, it appears likely that the conserved repeating segments incorporate the shared $Ca^{2+}$- and phospholipid-binding sites, whereas the various *N*-termini mediate the biological function(s) unique to each member. The annexins, characteristically, are water soluble, bind to acidic phospholipids in a $Ca^{2+}$-dependent manner, are amphipathic and lack an obvious

'EF-hand' $Ca^{2+}$-binding motif which identifies $Ca^{2+}$-binding proteins of the calmodulin/troponin C family. As a result, it appears possible that the annexins' $Ca^{2+}$-binding site arose independently from that of the aforementioned proteins.

Although the conservation of annexins during evolution argues strongly for important physiological roles, these roles remain confusing and controversial. The annexins have been variously ascribed to be anti-inflammatory via the inhibition of phospholipase $A_2$ (annexins I–VI), to inhibit blood coagulation (annexins I–VI), to mediate exocytosis via membrane–membrane fusion, to regulate cytoskeletal organization via actin bundling (annexins I and II) and to form voltage-gated cation channels with a high selectivity for $Ca^{2+}$ (annexins V and VII). In addition, annexin III has been reported to be identical to inositol-1, 2-cyclic phosphate 2-phosphohydrolase, an enzyme of the phosphotidylinositol signalling pathway, and annexin V has been claimed to mediate cell adhesion to collagen matrices. The capacities of these proteins to inhibit phospholipase $A_2$ is now widely considered to reflect the characteristic ability of the annexins to bind phospholipids (the substrate) in a $Ca^{2+}$-dependent manner, rather than to the enzyme *per se*. Indeed, substrate 'sequestration' represents the most satisfying explanation for their observed inhibition of phospholipase $A_2$ and blood coagulation. In this case, the latter activities represent *in vitro* artefacts rather than an accurate description of their *in vivo* physiological roles. In other words, to a large extent their functions remain a mystery and thus represent a considerable challenge for the future. The function(s) of many proteins are regulated by post-translational modifications, especially phosphorylation, and the identification of such changes and how they are induced has often provided valuable clues to the respective protein's function. In particular, *in vivo* phosphorylation of annexins I and II has been reported to be induced by stimulation of cell growth with epidermal growth factor and by viral transformation (i.e. $pp60^{v\text{-}src}$), respectively, and to profoundly affect their $Ca^{2+}$ requirement for phospholipid binding. Whilst these results are consistent with annexins I and II playing important roles in cell transformation and stimulation of cell growth, such roles have yet to be delineated.

A notable recent achievement has been the solving by X-ray diffraction of the three-dimensional crystal structure of annexin V at 2 Å resolution. This structure has, in turn, been related to that revealed by electron microscopy for two-dimensional crystals located on phospholipid monolayers. These studies have provided detailed information on overall shape (64 × 40 × 30 Å), the secondary and tertiary structure (four domains each with five α-helices), the nature of the $Ca^{2+}$-binding sites [Gly-Xaa-Gly-Thr-(38 residues)-Asp or Glu] and of the interaction of the annexins with phospholipids/membranes. Most importantly, it is evident that the tertiary arrangement of the polypeptide chain is unique, which suggests that the annexin supergene family arose separately during evolution. Additionally, these structural studies provide a rational basis to the possibility that annexin V can function as a $Ca^{2+}$-channel, by using a novel mechanism based on the generation of a local gradient of electrostatic potential at the protein–membrane interface.

As is so often the case, the present state of knowledge of the annexins poses yet more questions, the most over-riding of which concern biological function. In this context, the recent structural insights may provide an important clue to this function, especially in relation to their possible mediation of ion-conductance across membranes. However, the need for satisfying explanations for the diversity of structure at the *N*-terminus, for the roles of post-translational modifications and for their cellular, tissue and species distribution, still remain to entice new investigations and investigators.

**M. J. Crumpton**
*Imperial Cancer Research Fund 1992*

# Contents

# Abbreviations

| | |
|---|---|
| **ABC** | ATP-binding cassette |
| **CABP** | Cyclic AMP-binding protein |
| **CFTR** | Cystic fibrosis transmembrane regulator |
| **CPB** | Calcium phospholipid binding protein |
| **EF** | Elongation factor |
| **EGF** | Epidermal growth factor |
| **GRE** | Glucocorticoid response element |
| **ISA** | Intestinal-specific annexin |
| **NBF** | Nucleotide-binding fold |
| **NGF** | Nerve growth factor |
| **PCR** | Polymerase chain reaction |
| **PI** | Phosphatidylinositol |
| **$PLA_2$** | Phospholipase $A_2$ |
| **PLBP** | Phospholipid-binding protein |
| **RSV** | Rous sarcoma virus |
| **sGRE** | Slow glucocorticoid response element |
| **VAC** | Vascular anticoagulant |

# The annexins

**Stephen E. Moss**

Department of Physiology, University College, London WC1E 6BT, U.K.

## Introduction

Calcium has long been recognized as playing a central role in the regulation of many physiological processes. A dietary requirement for calcium has been acknowledged for more than a century, and the roles of calcium in muscle contraction and nervous stimulation have been studied for many years. However, it was the discovery that calcium could act as a second messenger in stimulus–response coupling that provided the impetus for much current research into calcium-binding proteins. Molecular and structural characterization has shown that the largest single category of calcium-binding proteins, including calmodulin and troponin C, bind calcium through a simple helix-loop-helix structure known as the 'E-F' hand [1]. However, over the past few years another major class of calcium-binding proteins has emerged, that bind calcium through an alternative, more complex structure.

The characterization of these proteins by molecular cloning has revealed a family of at least nine unique genes in mammals, two in *Drosophila* and one in *Hydra*. Various generic names have been used to describe the family, including annexins, lipocortins, calpactins and endonexins (see Table 1) [2,3]. A burgeoning literature testifies to increasing worldwide interest in the annexins, and accounts for the provocative but equivocal data suggesting that they function in a broad range of biological processes. This volume, devoted exclusively to the annexins, reviews their major proposed functions, including inhibition of phospholipase $A_2$ and of blood coagulation, membrane fusion and $Ca^{2+}$-channel activity.

## Defining an annexin

Although the mammalian annexin family comprises at least nine gene products, other annexins, some of which are clearly not homologues of the

**Table 1. Nomenclature of the annexin family**

| Annexin | Synonyms |
|---|---|
| I | Lipocortin I |
| | Calpactin II |
| | p35 |
| | Chromobindin 9 |
| II | Calpactin I heavy chain |
| | Lipocortin II |
| | p36 |
| | Chromobindin 8 |
| | Protein I |
| | Placental anticoagulant protein IV |
| III | Lipocortin III |
| | Placental anticoagulant protein III |
| | 35-$\alpha$ calcimedin |
| IV | Endonexin I |
| | Protein II |
| | 32.5 kDa calelectrin |
| | Lipocortin IV |
| | Chromobindin 4 |
| | Placental anticoagulant protein II |
| | Placental protein 4-X |
| | 35-$\beta$ calcimedin |
| V | Placental anticoagulant protein I |
| | Inhibitor of blood coagulation |
| | Lipocortin V |
| | 35 kDa calelectrin |
| | Endonexin II |
| | Placental protein 4 |
| | Vascular anticoagulant-$\alpha$ |
| | 35-$\gamma$ calcimedin |
| | Calphobindin I |
| | Anchorin CII |
| VI | p68, p70, 73k |
| | 67 kDa calelectrin |
| | Lipocortin VI |
| | Protein III |
| | Chromobindin 20 |
| | 67 kDa calcimedin |
| | Calphobindin II |
| VII | Synexin |
| VIII | Vascular anticoagulant-$\beta$ |
| IX and X | (only reported in *Drosophila*) |
| XI | |
| XII | (only reported in *Hydra*) |

*The recently cloned intestinal-specific annexin (ISA) is not included, but might tentatively be regarded as annexin XIII.*

**Fig. 1. The canonical annexin structure.**

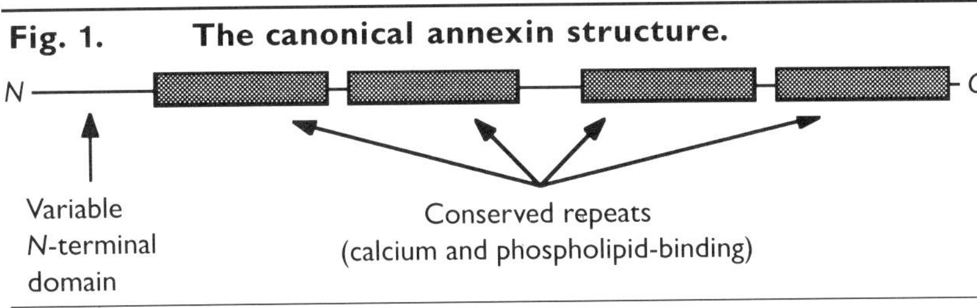


*A typical four-repeat annexin is shown, emphasizing the arrangement of the calcium/phospholipid-binding sites.*

**Fig. 2. Annexin protein structure.**

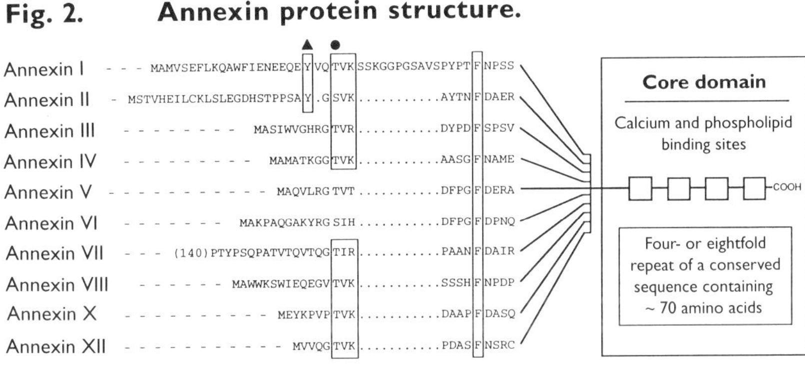


*The sequences of the N-terminal domains of the annexins show limited areas of sequence similarity (boxed). ▲, tyrosine kinase phosphorylation site; ●, probable protein kinase C phosphorylation site. Figure prepared by H. Haigler.*

known mammalian annexins, have been identified in various lower phyla, including *Drosophila* [4], *Hydra* [5], *Dictyostelium* [6] and higher plants [7]. The annexins are thus broadly expressed, both phylogenetically and within cells and tissues. For a protein to qualify as a member of the annexin family, it must satisfy two essential criteria, one structural and one biochemical. Respectively, these are the presence of a conserved 70-amino-acid domain, repeated either four or eight times in the overall structure, and the ability to bind calcium-dependently to phospholipids. For all annexins, the conserved repeats account for the major part or 'core' of the protein, and it is these regions that harbour the calcium and phospholipid-binding sites (Fig. 1).

At the *N*-terminus of each annexin is a unique domain, varying both in length and amino-acid composition. Given that the core domains are broadly similar for all annexins, it seems likely that the *N*-termini confer functional individuality [8]. This is illustrated in Figs. 2 and 3, which show that the *N*-termini of annexins I and II contain various unique features, including sites for phosphorylation by tyrosine and serine/threonine kinases (see chapters 5 and 2 by Gerke, and Haigler &

**Fig. 3. Comparison of the *N*-termini of annexins I and II.**

**Annexin I**

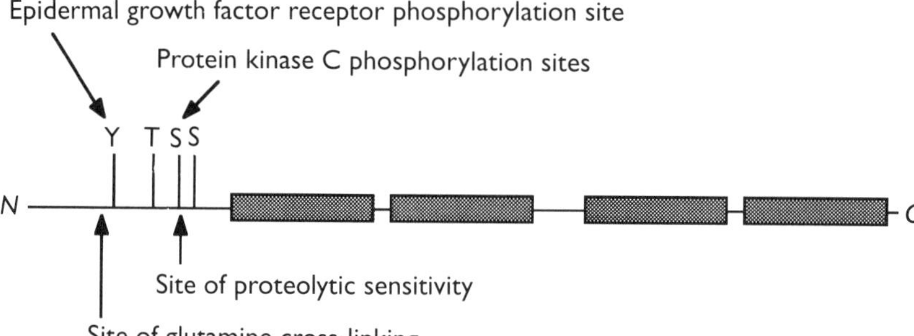


**Annexin II**

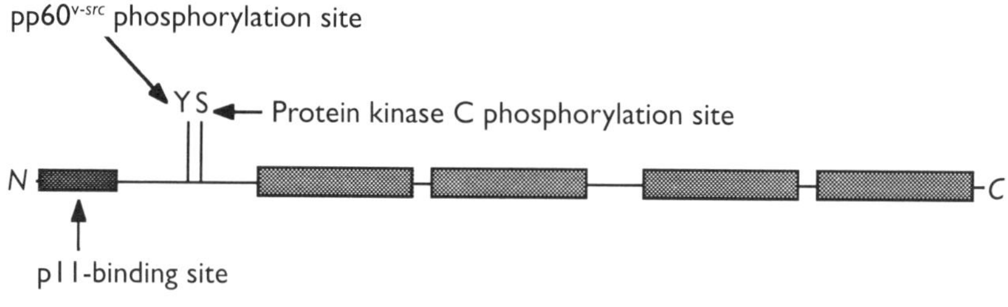


*The N-terminal domains of annexins I and II are similar in that they contain phosphorylation sites, but differ in other ways.*

Schlaepfer, respectively), and in the case of annexin II, the binding site for a cellular ligand, p11 (see chapter 6 by Weber). In addition to *N*-terminal variation, alternative splicing may generate further functional diversity for certain members of the annexin family. Both annexins VI and VII have alternatively spliced exons within their coding sequences (see Fig. 4). For annexin VI, the inclusion of this exon results in a six-amino-acid insert at the start of the seventh repeat [9], whereas for annexin VII the alternative splice site lies within an unusually long and hydrophobic *N*-terminus (see chapter 9 by Pollard *et al.*).

## The annexin genes

The discovery of annexins in lower eukaryotes suggests that in evolutionary terms the origins of the family must indeed be ancient. This

**Fig. 4. Alternative splicing of annexins VI and VII.**

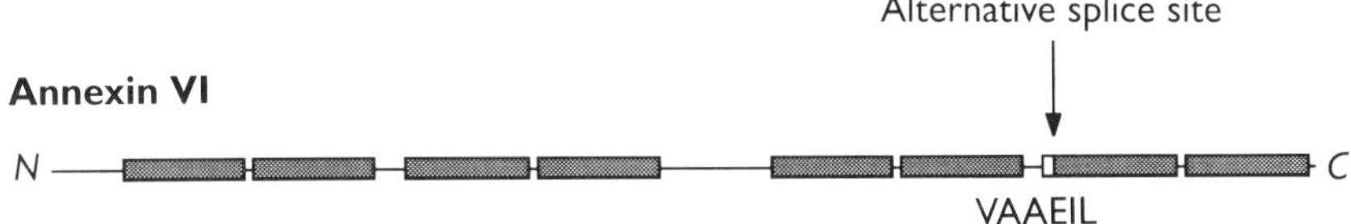


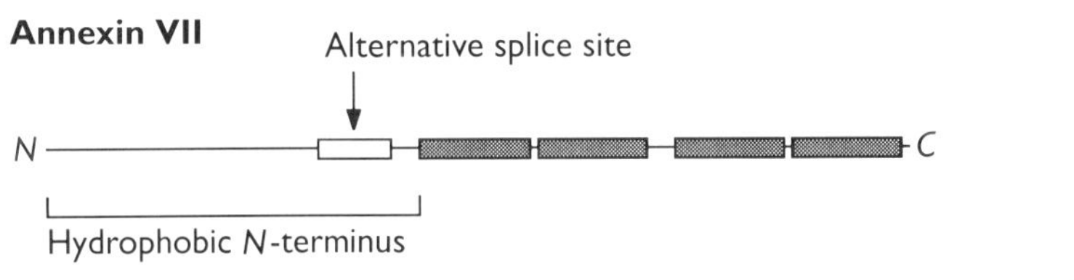


*Alternative splice sites exist at the start of the seventh repeat of annexin VI, and within the* N*-terminal domain of annexin VII.*

idea is consistent with the fact that the chromosomal loci of the human annexin genes are widely distributed within the genome (Table 2). Several groups have recently started to analyse the structures of the annexin genes, and in doing so have revealed some surprising results (see chapter 3 by Horseman) [10–14]. It might have been expected that the repetitive structure of the annexins observed at the amino-acid level would be reflected in the organization of the exons. However, the human annexin I and II genes display a variety of intron–exon boundaries within the four repeats of the core domain. Not only do exons span repeats, but intron–exon boundaries also lie within the $Ca^{2+}$-binding sites; this is analogous to the presence of introns found within the $Ca^{2+}$-binding loops of certain E-F hand proteins.

However, the positions of these boundaries are extremely well conserved between the corresponding repeats of the two proteins,

**Table 2. Human annexin gene loci**

| Annexin | Chromosome |
|---|---|
| I | 9q |
| II | 15q (4, 9, 10) |
| III | 4q |
| IV | 2 |
| V | 4q |
| VI | 5q |
| VII | 10q |
| VIII | 10q |

*The loci in parentheses for annexin II indicate pseudogenes.*

suggesting that the evolution of the tetrad repeat preceded the divergence of the two annexins. Our preliminary studies on the annexin VI gene (P. D. Smith, M. J. Crumpton & S. E. Moss, unpublished work) show that the positions of the intron–exon boundaries are virtually all different from those of annexins I and II. This further suggests that annexin VI may have diverged prior to the split between annexins I and II. Clearly, a full analysis of all the human annexin genes will provide valuable insights into the evolution of this family.

Certain annexins may also exist as sub-families, with two or more distinct genes encoding virtually identical proteins. This was first observed for pigeon annexin I [15] and has been reported more recently for *Xenopus* annexin II [16]. The most intriguing aspect of these observations is that these phenomena appear to be species specific. Thus, there is no evidence that either annexins I and II have more than one functional gene in humans, although it is interesting to note that there are three human annexin II pseudogenes.

Cloning of the annexin genes and the availability of their promoter sequences will accelerate our elucidation of the molecular basis of their tissue specificity. Although representatives of the annexin family are expressed in all cell types, the combination of annexins varies from one tissue to another. As individuals, most of the annexins are actually quite restricted in their tissue distribution. For example, annexin VI (see chapter 11 by Dedman & Kaetzel) is strongly expressed in the islet cells of the pancreas but not in the exocrine cells [17]. The expression of other annexins may be growth-dependently [18] or developmentally [17,19,20] regulated. Perhaps the most striking example of tissue specificity is the recently cloned intestinal-specific annexin (ISA) which is expressed only in intestinal epithelial cells [21]. The differential expression of annexins indicates that, despite their similar structures, they have quite distinct modes of gene regulation.

## The problems of function

The accumulation of structural and biochemical data on the annexins has progressed far more rapidly than our understanding of their functions. Although most would accept that no single annexin has been unequivocally associated with a particular function, several lines of investigation have persisted in the literature. Most provocative is the suggestion that annexin I (lipocortin) is a mediator of anti-inflammatory responses, acting through the inhibition of phospholipase $A_2$ (see chapter 4 by Russo-Marie) [22–24]. More recently, annexins have also been described as inhibitors of blood coagulation (see chapters 13 and 12 by Maki *et al.* and Hauptmann & Reutelingsperger, respectively) [25–27], although both this phenomenon and phospholipase inhibition can be accounted for *in vitro* by calcium-dependent sequestration of phospholipid [28,29]. One of the more attractive functional theories suggests a link between certain annexins and secretion (see chapters 8 and 7 by Creutz *et al.* and Burgoyne, respectively). Annexin II (calpactin) stimulates aggregation of isolated chromaffin granules at physiological concentrations of $Ca^{2+}$ [30], and prevents 'run down' of exocytotic capacity when added to permeabilized cells [31].

Attention has recently focused on annexin V, the first of the family to be structurally characterized by X-ray crystallography (see chapter 10 by Huber *et al.*) [32]. The structure of annexin V underlies the theory that annexin V functions as a $Ca^{2+}$ channel. Importantly, this suggestion is supported by electrophysiological data showing that annexin V exhibits classic voltage-gated $Ca^{2+}$-channel activity in artificial lipid bilayers [33]. However, the $Ca^{2+}$-channel activity of annexin V (and annexin VII which also has this property) has still to be demonstrated in whole cells; alternative functional theories exist, the most intriguing of which is that annexin V is a specific and potent inhibitor of protein kinase C [34].

There is no question that many of the functional theories suggested for the annexins are based on compelling biochemical and physiological data. It is perhaps too easy to be dismissive of some of these ideas, for example phospholipase $A_2$ inhibition, simply because they can be explained by $Ca^{2+}$-dependent phospholipid sequestration. If annexin I does have an anti-inflammatory role, then this may be a perfectly reasonable mechanism of action. However, all annexins inhibit phospholipase $A_2$ *in vitro*, and it seems unlikely that the members of such a diverse gene family could share the same function *in vivo*.

Clearly, the core domain common to the annexins must imply at least some common functional features. But it may be that this domain (which harbours the $Ca^{2+}$- and phospholipid-binding sites) simply serves a subcellular targetting role, directing the protein to the correct compartment following synthesis. This idea is consistent with the *N*-terminal domains being responsible for defining the individual functions of the annexins. With so many proposed functions it is unreasonable to suppose that all are correct, but it is equally hard to imagine that all are incorrect. Whatever the answers, the search for function in the annexin family is likely to enjoy a healthy level of controversy for several years to come.

## References


1. Tufty, R. M. & Kretsinger, R. H. (1975) Science **187**, 167–169
2. Crumpton, M. J. & Dedman, J. R. (1990) Nature (London) **345**, 212
3. Browning, J., Pepinsky, B., Wallner, B., Flower, R. J. & Peers, S. H. (1990) Nature (London) **346**, 324
4. Johnston, P. A., Perin, M. S., Reynolds, G. A., Wasserman, S. A. & Sudhof, T. C. (1990) J. Biol. Chem. **265**, 11382–11388
5. Schlaepfer, D. D., Fisher, D. A., Bode, H. R., Jones, J. M. & Haigler, H. T. (1991) J. Cell Biochem. **S15B**, 218
6. Doring, V., Schleicher, M. & Noegel, A. A. (1991) J. Biol. Chem. **266**, 17509–17515
7. Smallwood, M., Keen, J. N. & Bowles, D. J. (1990) Biochem. J. **270**, 157–161
8. Crompton, M. R., Moss, S. E. & Crumpton, M. J. (1988) Cell **55**, 1–3
9. Moss, S. E. & Crumpton, M. J. (1990) FEBS Lett. **261**, 299–302
10. Hitti, Y. S. & Horseman, N. D. (1991) Gene **103**, 185–192
11. Spano, F., Raugei, G., Palla, E., Colella, C. & Melli, M. (1990) Gene **95**, 243–251
12. Amiguet, P., D'Eustachio, P., Kristensen, T., Wetsel, R. A., Saris, C. J. M., Hunter, T., Chaplin, D. D. & Tack, B. F. (1990) Biochemistry **29**, 1226–1232
13. Kovacic, R. T., Tizard, R., Cate, R. L., Frey, A. Z. & Wallner, B. P. (1991) Biochemistry **30**, 9015–9021
14. Horlick, K. R., Cheng, I. C., Wong, W. T., Wakeland, E. K. & Nick, H. S. (1991) Genomics **10**, 365–374
15. Horseman, N. D. (1989) Mol. Endocrinol. **3**, 773–779
16. Gerke, V., Koch, W. & Thiel, C. (1991) Gene **104**, 259–264
17. Clark, D. M., Moss, S. E., Wright, N. A. & Crumpton, M. J. (1991) Histochemistry **96**, 405–412
18. Keutzer, J. C. & Hirschhorn, R. R. (1990) Exp. Cell Res. **188**, 153–159
19. McKanna, J. A. & Cohen, S. (1989) Science **243**, 1477–1479
20. Carter, V. C., Howlett, A. R., Martin, G. S. & Bissel, M. J. (1986) J. Cell Biol. **103**, 2017–2024
21. Wice, B. M. & Gordon, J. I. (1992) J. Cell Biol. **116**, 405–422
22. Whitehouse, B. J. (1989) J. Endocrinol. **123**, 363–366
23. Flower, R. J. (1988) Br. J. Pharmacol. **94**, 987–1015
24. Cirino, G. & Flower, R. J. (1991) in Novel Calcium-binding Proteins (Heizmann, C. W., ed.), pp. 589–611, Springer-Verlag, Heidelberg
25. Tait, J. F., Sakata, M., McMullen, B. A., Miao, C. H., Funakoshi, T., Hendrickson, L. E. & Fujikawa, K. (1988) Biochemistry **27**, 6268–6276
26. Funakoshi, T., Heimark, H., Hendrickson, L. E., McMullen, B. A. & Fujikawa, K. (1987) Biochemistry **26**, 5572–5578
27. Yoshizaki, H., Mizoguchi, T., Arai, K., Shiratsuchi, M., Shidara, Y. & Maki, M. (1990) J. Biochem. (Tokyo) **107**, 43–50
28. Davidson, F. F., Dennis, E. A., Powell, M. & Glenney, J. R. (1987) J. Biol. Chem. **262**, 1698–1705

29. Edwards, H. C. & Crumpton, M. J. (1991) Eur. J. Biochem. **198**, 121–129
30. Drust, D. S. & Creutz, C. E. (1988) Nature (London) **331**, 88–91
31. Ali, S. M., Geisow, M. J. & Burgoyne, R. D. (1989) Nature (London) **340**, 313–315
32. Huber, R., Romisch, J. & Paques, E. P. (1990) EMBO J. **12**, 3867–3874
33. Rojas, E., Pollard, H. B., Haigler, H. T., Parra, C. & Burns, A. L. (1990) J. Biol. Chem. **265**, 21207–21215
34. Schlaepfer, D. D., Jones, J. & Haigler, H. T. (1992) Biochemistry **31**, 1886–1891

# Annexin I phosphorylation and secretion

**Harry T. Haigler and David D. Schlaepfer**

Department of Physiology and Biophysics, California College of Medicine, The University of California, Irvine, CA 92717, U.S.A.

Annexin I earned its name by being the first annexin to have its amino acid sequence revealed. For reasons that extend beyond its name, it still remains at the forefront of the annexin field in attracting active investigation. Annexin I is interesting because it is a substrate for two different kinases that play key roles in intracellular signal transduction. In addition, it has the intriguing property of being both a cytosolic and a secreted protein. But perhaps the most interesting aspect of annexin I is that it enjoys great controversy. Despite extensive studies, its biological function is not yet known. Thus, the physiological significance of annexin I phosphorylation is difficult to interpret. It is also difficult to reconcile the observed secretion with the knowledge that annexin I cDNA does not predict a hydrophobic signal sequence to direct it to the classic cellular secretory pathway.

This chapter will review the studies of annexin I phosphorylation by the epidermal growth factor (EGF) receptor/kinase and protein kinase C. It will also review the work on annexin I secretion. The reader should be aware that phosphorylation and secretion are two separate topics and there is no indication that they are related events.

## Overview of annexin I phosphorylation

Phosphorylation of key regulatory proteins is one of the primary mechanisms by which cells respond to extracellular signals. Over the past 10 years a class of protein kinases that phosphorylate tyrosine residues has attracted intense investigation [1]. This interest arises largely from the involvement of protein-tyrosine kinases in the normal processes of growth and differentiation and in abnormal growth associated with neoplastic transformation. Several growth factor receptors are transmembrane proteins that have a tyrosine kinase activity as part of their cytosolic domains [2]. Non-receptor protein-tyrosine kinases such as the proto-oncogene pp60$^{src}$ also generate growth-promoting signals. Like the catalytic domain of the growth factor receptors, the non-receptor protein-tyrosine kinases are often localized to the cytosolic face of the plasma membrane [1]. Although rapid advances have been made in understanding the structure and function of these kinases, it has been more difficult to identify their physiological cellular substrates and to determine how phosphorylation affects their biological function. In the past few years however, a small number of cellular substrates have been identified including annexins I and II.

Annexin I and annexin II (see chapter 5 by Gerke) are excellent *in vitro* substrates for the EGF receptor/kinase and for pp60$^{src}$, respectively. Although the phosphorylated tyrosines in both annexins are located at analogous positions in their primary structures (see discussion below), the kinases demonstrate a high degree of substrate specificity; i.e. pp60$^{src}$ will not phosphorylate annexin I, and *vice versa*. Because annexins undergo reversible $Ca^{2+}$-dependent binding to phospholipids that are preferentially located on the cytosolic face of the plasma membrane [3,4], and because this is the location of these kinases, an attractive proposition is that annexins I and II are physiological substrates for these protein-tyrosine kinases. The properties of annexin I phosphorylation by the EGF receptor/kinase will be reviewed in detail below.

Several of the annexins, including annexin I, are also excellent substrates for protein kinase C *in vitro*. Protein kinase C is an important element in signal transduction pathways that can be stimulated by a number of extracellular effectors including some that promote growth [5]. Cellular protein kinase C is activated by $Ca^{2+}$-dependent binding of the kinase to phosphatidylserine and diacylglycerol on the cytosolic face of the plasma membrane. Here again, the juxtaposition of the annexin and the kinase at the same cellular location makes it tempting to speculate about the possible biological functions of protein kinase C phosphorylation of annexin I. This topic will also be covered below.

It is informative to consider annexin phosphorylation in light of the common structural features of this family of proteins. Each annexin has two domains: a small *N*-terminal domain and a larger core domain that contains the $Ca^{2+}$- and phospholipid-binding activities that are common to all annexins [4]. The core domains share approximately 50% amino acid sequence identity, whereas their *N*-terminal domains have quite diverse sequences. All known phosphorylation sites of annexins are located in their *N*-terminal domains. Even though the annexin sequences are divergent in the *N*-terminus, protein kinase C recognition sites with similar structural features are present at analogous positions in this domain of most annexins. The protein-tyrosine kinase sites found in annexins I and II are located at analogous positions a few residues upstream of the protein kinase C phosphorylation sites in these two proteins. Because the structures of the *N*-terminal domains of the annexins are quite divergent in general, it seems likely that the similarities observed in phosphorylation result from convergent evolution, rather than arising from a common ancestral gene.

## Phosphorylation of annexin I by the EGF receptor/kinase

The first indication that annexin I was a substrate for the EGF receptor/kinase came from studies by Fava and Cohen [6] on the A431 carcinoma cell line which over-expresses the EGF receptor. They identified a protein of apparent $M_r$ 35000 that was phosphorylated on tyrosine in an EGF-dependent manner by A431 membranes in an *in vitro* assay. Fava and Cohen [6] noted that the 35 kDa substrate underwent reversible $Ca^{2+}$-dependent binding to membranes, and exploited this observation in the key step of a purification protocol that allowed isolation of the protein. The basic outline of their isolation procedure has been used or independently discovered by several other laboratories and has been central to the identification and purification of other annexins. A structural framework for investigating the phosphorylation of the 35 kDa substrate was provided with the elucidation of the sequence of lipocortin I (later named annexin I) [7] and the recognition that this protein was identical to the 35 kDa substrate [8–11]. Annexin I is phosphorylated by the EGF receptor/kinase at Tyr-21 [8,10]. This phosphorylation site of annexin I, which is preceded by several acidic residues, resembles the sites of tyrosine phosphorylation in certain other protein substrates including the major autophosphorylation site of the EGF receptor/kinase.

An important feature of the EGF receptor/kinase phosphorylation of annexin I is that the reaction is of high affinity and is $Ca^{2+}$-dependent. Annexin I was a very high-affinity substrate for the EGF-stimulated kinase in A431 membranes; half-maximal phosphorylation occurred at an annexin I concentration of 50 nM [10]. Because annexins bind to phospholipids with high affinity (association constants in the pM range [12]), this property may contribute to the high affinity of the phosphorylation reaction. This proposal is strengthened when the role played by $Ca^{2+}$ in the phosphorylation reaction is considered. When the reaction was performed in the presence of $Mg^{2+}$ as the co-factor for the EGF receptor/kinase in A431 membranes, annexin I phosphorylation was absolutely dependent on $Ca^{2+}$ even though $Ca^{2+}$ had no affect on EGF receptor/kinase autophosphorylation [6,10]. The $Ca^{2+}$ concentration required for either half-maximal phosphorylation or binding to phospholipid vesicles was very similar—28 μM and 22 μM [3], respectively. Thus, annexin I binding to the phospholipids on the cytosolic face of the A431 membranes may have been an essential step in positioning it next to the kinase and resulting in high-affinity phosphorylation. Additional support for this proposal comes from phosphorylation studies with cations other than $Ca^{2+}$. If the A431 membrane phosphorylation reaction contained $Mn^{2+}$, $Ca^{2+}$ was not required for annexin I phosphorylation. Additionally, it was shown that $Mn^{2+}$, but not $Mg^{2+}$, would support annexin I binding to phospholipid vesicles [3]. The phosphorylation of annexin I was also observed in reactions containing $Mg^{2+}$ and cations that support annexin I binding to phospholipid, such as $Sr^{2+}$ or $Ba^{2+}$ (J. Fitch & H. Haigler, unpublished work).

Thus, several separate lines of evidence suggest indirectly that the high-affinity phosphorylation of annexin I observed *in vitro* is secondary to $Ca^{2+}$-dependent binding of the substrate to the phospholipids on the membrane. This topic needs further investigation with purified detergent-solubilized EGF receptor/kinase. It will be particularly interesting to determine whether $Ca^{2+}$-dependent relocation of annexin I from the cytosol to the plasma membrane is necessary for annexin I phosphorylation in intact cells.

It is worth noting that a form of annexin I has been observed in human placenta that associates with membranes in a $Ca^{2+}$-independent manner [11,13]. The mechanism of attachment is not known. However, it is clear that the endogenous EGF receptor/kinase does not require $Ca^{2+}$ for phosphorylation of this form of annexin I [11,13].

Several studies in intact cells indicate that tyrosine phosphorylation of annexin I is a biologically important event and not an artifact that is confined to test tube assays. Annexin I was phosphorylated in intact A431 carcinoma cells and the extent of phosphorylation was increased up to 100-fold by treatment of cells with EGF [14]. The majority of phosphate was incorporated into phosphotyrosine as would be expected for EGF receptor/kinase-catalysed phosphorylation. However, some phosphate was also incorporated into phosphoserine, thereby raising the possibility of interaction with other kinases such as protein kinase C. Annexin I phosphorylation was also stimulated approximately 10-fold by EGF in normal diploid human fibroblasts that have a well defined growth response to the hormone [15]. Only phosphotyrosine was observed in EGF-treated intact fibroblasts. Maximal EGF-dependent annexin I phosphorylation occurred at 1 h in intact fibroblasts [15]. Annexin I was shown *in vitro* to associate with and be phosphorylated by endosomes that contain the internalized EGF receptor/kinase [16]. As the time courses of both endosome internalization and annexin I phosphorylation are similar, it has been proposed that annexin I is involved with the endocytosis or vesicular trafficking associated with EGF internalization and degradation [16].

Before a protein can be seriously considered as a physiological substrate for a cellular kinase, it is important to determine whether it is phosphorylated to a stoichiometry that could reasonably be expected to have a significant effect on the activity of the protein within the cell. Preliminary experiments in normal human fibroblasts indicate that the stoichiometry of annexin I phosphorylation on Tyr-21 increased from less than 1% to 20% when quiescent cells were stimulated to grow with fresh serum and EGF [17]. These results are consistent with phosphorylation having a biologically important impact on the function of this protein.

The majority, but not all, of the data collected to date indicate that phosphorylation of annexin I is a biologically important event. However, some studies present evidence that this phosphorylation event may not be directly related to mitogenic signal transduction. First, annexin I is expressed in certain cell types that do not respond to growth factor stimulation. For example, cells in the neural tube of embryonic mice begin to express high levels of annexin I at just the time that they undergo terminal differentiation into non-dividing cells [18]. Second, it is curious that phosphorylated annexin I was not detected by several different groups that were searching for substrates for protein-tyrosine kinases using an antibody that recognizes phosphotyrosine residues. However, lack of detection by these methods may be because of technical problems. There is concern that the anti-phosphotyrosine antibody used in most of these studies will not recognize annexin I phosphorylated on Tyr-21. In addition, it has been shown previously that phosphorylation of Tyr-21 renders annexin I 20-fold more sensitive to a proteolytic clip which removed the *N*-terminal 26 amino acids that contained the phosphorylated tyrosine residue [10]. It is not known what, if any, physiological significance is associated with this increased proteolytic sensitivity, but it is possible that phosphorylated annexin I was selectively degraded by proteases in cellular extracts before the analysis was completed.

In summary, some studies indicate that phosphorylation of annexin I is involved in mitogenic signal transduction, whereas other studies are inconsistent with this role. When faced with the confusing array of data in this field, it might be wise to consider the possibility that annexin I is performing different functions in different tissues and cell types. It is well established that the $Ca^{2+}$-binding protein calmodulin regulates different biological functions in different cell types by interacting with different regulatory proteins. Thus, one should maintain an open mind when studies in different biological systems produce apparently conflicting data concerning annexin function.

## Phosphorylation of annexin I by protein kinase C

Annexin I is a high-affinity substrate for protein kinase C *in vitro* [19–21] and is phosphorylated in a stimulant-dependent manner in intact chromaffin cells [22]. *In vitro* phosphorylation of human annexin I occurs to an equivalent extent on Thr-24, Ser-27 and Ser-28 [21] and stoichiometry of phosphorylation of greater than 1 was observed. It is interesting to note that guinea pig annexin I has a structure identical to human annexin I at and around the three phosphorylated residues, whereas annexin I from rat, mouse, pigeon (cp37) and sponge is conserved only at the Ser-27 phosphorylation site. Because protein kinase C phosphorylation of annexin I probably plays similar roles in all of these species, one might suspect that phosphorylation of Thr-24, Ser-27 or Ser-28 of human annexin I would have equivalent biological effects.

Although annexin I is an excellent substrate for protein kinase C *in vitro*, the physiological significance of this modification is not yet known. Only limited studies have been performed on intact cells to date. One study was unable to demonstrate phosphorylation of annexin I in phorbol ester-treated A431 cells [23], but annexin I was phosphorylated on serine in EGF-treated A431 cells [14]. Because a protein kinase C phosphorylation site on annexin I has been conserved from humans to sponges, it is likely that this serves a biologically important function.

A recent study showed that annexin I phosphorylation by protein kinase C was inhibited *in vitro* by annexin V [24]. Inhibition apparently occurred by direct interaction between the kinase and annexin V. Half-maximal inhibition occurred at an annexin V concentration comparable to that found in many cells, thereby indicating that the *in vitro* data may correspond to a physiologically important event. Because annexin V has a cellular half-life of over 20 h [25], it may not participate in acute regulation of protein kinase C activity. But it is interesting to note that annexin V levels decreased about four-fold as quiescent human fibroblasts entered a period of rapid growth, and annexin I levels increased four-fold during the same period [25]. Perhaps annexin V exerts a chronic effect on protein kinase C activity during the relatively slow changes involved in growth regulation?

## Phosphorylation summary

For reasons outlined in the preceding sections, it is reasonable to propose that phosphorylation of annexin I is a physiologically important event. Additional support for this proposal comes from the observation that two different forms of annexin I, cp35 and cp37, are expressed in the pigeon crop sac (see chapter 3 by Horseman). These two proteins share 93% overall amino acid sequence identity but have quite different sequences in a short segment of the *N*-terminal domain where phosphorylation sites for the EGF receptor/kinase and protein kinase C are located. The fact that evolution selected for two annexin I-like proteins, in which the only major structural differences are located in and around the phosphorylation sites, indicates that this region plays an important role in regulation of the biological activity of the protein.

The most important problem regarding annexin I phosphorylation is to determine how these modifications affect the protein's biological activity. Unfortunately, this question cannot be addressed directly because the biological activity of annexin I is not yet known. So far, only two effects of phosphorylation have been noted. First, phosphorylation of Tyr-21 results in a 20-fold increased sensitivity to a proteolytic clip at Lys-26 [10]. It is possible that the phosphorylated peptide is the biologically active species after it is clipped from the protein and the $Ca^{2+}$-dependent phospholipid binding activity associated with the core domain only serves to direct the protein to the plasma membrane for phosphorylation. Alternatively, the function of the *N*-terminal domain may be to modify the activity of the core domain. Other studies showing that proteolytic removal of the *N*-terminal domain of annexin I increased the affinity of the core domain for $Ca^{2+}$ [26] are consistent with the latter alternative. Second, phosphorylation of annexin I at Tyr-21 resulted in a five-fold decrease in the concentration of $Ca^{2+}$ required to promote annexin I binding to phospholipid [3]. These observations support the proposal that the *N*-terminal domain can modify the activity of the core domain.

## Secretion of annexins

Proteins destined for secretion by eukaryotic cells usually contain a 'signal sequence' that directs them to the lumen of the endoplasmic reticulum [27]. The signal sequence is usually located on the *N*-terminus of the nascent peptide chain and is cleaved off during translocation [27]. Although a number of unanswered questions remain concerning the molecular mechanism by which targeting and translocation occurs, the hypothesis that the signal sequence directs a protein to the secretory pathway has been very successful in predicting the cellular fate of eukaryotic, prokaryotic and fusion proteins.

However, interleukin 1 [28,29], fibroblast growth factor [30,31] and a few other proteins [32] that are thought to have extracellular functions lack signal sequences. Recent studies, particularly those on interleukin 1 [33–36], have provided reasonably convincing data for the existence of a selective unconventional secretion mechanism. Certain annexins may also be secreted by unconventional pathways even though they do not have signal sequences. Studies reviewed below clearly show that annexin I is selectively secreted by the prostate gland and perhaps also by other cells.

It is quite clear that all annexins are abundant intracellular proteins in many cell types. The cDNA sequences of known annexins do not predict signal sequences for secretion, and all immunolocalization studies to date have detected annexins within the cytoplasmic compartment and have not provided evidence for their presence within secretory organelles [37–40]. Annexins bind in a reversible $Ca^{2+}$-dependent manner to phospholipids that are preferentially localized to the cytoplasmic face of cell membranes [3,41,42] and have the potential for involvement in $Ca^{2+}$-mediated stimulus–response coupling. The evidence for an intracellular location is particularly compelling for annexin I. It has an acetylated *N*-terminus [43] (the enzyme that catalyses acetylations is known to be cytoplasmic [44]), lacks glycosylation and is phosphorylated by intracellular kinases. Furthermore, annexin I was not sequestered in microsomes in a cell-free translation system [45], nor was it secreted by *Xenopus* oocytes that were injected with transcripts encoding annexin I [45].

Despite the overwhelming evidence that annexins are primarily intracellular proteins that are not secreted by the classic secretory pathway, the topic of extracellular annexins has attracted active investigation motivated, in part, by the controversial proposal that extracellular annexins (also known as lipocortins) are involved in mediation of the anti-inflammatory actions of steroids. This review will focus on the more recent studies in which the annexins were structurally defined. The original studies of the lipocortins are reviewed in chapter 4 by Russo-Marie.

The Biogen group showed that annexins I, III and V, but not II, IV and VI, were present in low concentrations (approximately 0.5 μg/ml) in the peritoneal exudates of dexamethasone-treated rats [46,47]. In fact, the first amino acid sequence data were derived from annexin I purified from this source [46]. Annexins were also reported to be present in low concentrations in bronchoalveolar lavage fluid [48,49], serum [50] and amniotic fluid [50]. However, because the levels detected were quite low, and as several groups reported that annexins were not secreted from cultured cells [51–53], the idea of the selective secretion of annexins was slow to be accepted. Because the concept of the classic secretory pathway was so firmly established, many investigators initially assumed that annexins reached the extracellular space by a non-specific mechanism such as release from dead or damaged cells.

In order to address the question of whether annexin I secretion is a physiological process, extracellular fluids from a number of sources were systematically screened for the presence of annexin I [54]. Prostate fluid contained remarkably high concentrations of intact annexin I (60 μg/ml) and the secreted protein had identical biochemical and functional properties as the intracellular protein [54]. It also contained a truncated form, des-1-29-annexin I (80 μg/ml), that apparently was formed by a proteolytic clip during the process of secretion [54]. Prostate tissue contained high concentrations of intact annexin I (0.6 % of total cellular protein) but no truncated forms.

Annexin IV provided a very informative internal control for establishing the specificity of annexin I secretion. It was expressed in prostate tissue at the same high levels as annexin I, and immunofluorescent double-labelling experiments showed that both annexins IV and I co-localized to the same ductal epithelium cells of the prostate. Even though annexins IV and I were expressed in the same cells, annexin IV was not secreted; less than 0.1 μg/ml annexin IV was detected in prostate fluid. Control experiments showed that annexin IV was not secreted and then subsequently degraded. This study showed that selective secretion of annexin I occurred in the prostate gland and the results indicated that only certain members of the annexin family possess a necessary sorting signal to direct their secretion. It was also shown that annexin V was secreted by the prostate gland. The limited characterization of the prostate gland showing secretion of annexins I and V but not IV is so far consistent with previous studies showing that annexins I, III and V, but not II, IV and VI, were present in peritoneal exudates.

Most secretory cells of the prostate have an ultrastructure similar to other exocrine cells and contain a well developed endoplasmic reticulum, Golgi apparatus and secretory granules [55–57]. The primary transcripts of the major secretory proteins of the prostate contain signal sequences for secretion via this conventional pathway [58–60]. Because annexin I clearly does not have a signal sequence, it appears that it is secreted by a novel secretory pathway in the prostate. These studies in the prostate raise broader questions. Can annexin I be secreted in other locations and should extracellular functions be considered for this protein? The fact that no unusual cellular secretory machinery has been identified in the prostate, combined with previous observations that annexin I is located in lung lavage fluid [48,49] and in peritoneal exudates [46,47], indicates that annexin I secretion may occur in a number of locations. Studies on the secretion of other proteins such as interleukin 1 which also lack a signal sequence strengthen the proposal that proteins such as annexin I, which lack signal sequences, can be secreted by a number of different cell types.

Although direct experimental information is not yet available, it is interesting to speculate on the cellular pathway by which annexin I might be secreted. Because annexin I secreted into the seminal plasma has a blocked *N*-terminus, and as *N*-terminal acetylation of proteins occurs in the cytoplasm, annexin I probably passed through this compartment prior to secretion. Two conceptually different mechanisms could account for the process by which annexin I is secreted. First, annexin I could be concentrated in envaginations of plasma membrane which pinch off to form extracellular vesicles that subsequently lyse and release their contents into the extracellular fluid. There is some general precedence for this proposal because vesiculation of the plasma membrane followed by vesicle lysis was recently proposed to account for the externalization of a cytoplasmic lectin lacking a signal [61]. There is also some precedence for incorporation of annexin I into extracellular vesicles: vesiculation of chondrocyte plasma membranes appears to be responsible for the formation of bone matrix vesicles that are enriched in certain cytoplasmic proteins, including annexin I [62,63]. However, these vesicles appear to be stable long-lived structures that are involved in bone matrix formation [62] and there is no evidence for the subsequent release of annexin I from these vesicles into the extracellular fluid [63]. In prostate fluid, annexin I remained in the supernatant following high-speed centrifugation in either the presence or absence of EGTA [54]. This indicates that annexin I was not entrapped in vesicles or bound to membrane phospholipids. Thus, annexin I was probably not secreted from the prostate via membrane blebes.

The second possible mechanism of secretion would require annexin I to cross a cellular membrane. It could either cross the plasma membrane directly or it could enter a vesicular secretory pathway, either the conventional pathway or a novel pathway composed of uncharacterized vesicles. As tightly folded proteins appear to be unable to cross cellular membranes, other proteins, perhaps of the chaperone type [64], would be required during the translocation process. Mammalian multi-drug resistance transporters are membrane proteins whose physiological substrates have not yet been determined. Annexin I and other proteins without signal sequences are intriguing candidates for substrates of this family of transporters.

In summary, the clear demonstration that annexin I is selectively secreted by the prostate lends credence to the idea that the low concentrations of the protein observed in other extracellular fluids is the result of selective secretion. Because some annexins are secreted while others are not, structure–function studies of these structurally related proteins may provide a useful model system for determining the structural elements that direct the secretion of proteins lacking conventional signal sequences. Finally, the important question concerning the biological function of annexin I is further complicated by the detection of annexin I in extracellular fluids. We are how required to consider both intracellular and extracellular functions for this intriguing protein.

## References


1. Hunter, T. (1989) Curr. Opin. Cell Biol. **1**, 1168–1181
2. Ullrich, A. & Schlessinger, J. (1990) Cell **61**, 203–212
3. Schlaepfer, D. D. & Haigler, H. T. (1987) J. Biol. Chem. **262**, 6931–6937
4. Crompton, M. R., Moss, S. E. & Crumpton, M. J. (1988) Cell **55**, 1–3
5. Kikkawa, U. & Nishizuka, Y. (1986) Annu. Rev. Cell Biol. **2**, 149–178
6. Fava, R. A. & Cohen, S. (1984) J. Biol. Chem. **259**, 2636–2645
7. Wallner, B. P., Mattaliano, R. J., Hession, C., Cate, R. L., Tizard, R., Sinclair, L. K., Foeller, C., Chow, E. P., Browning, J. L., Ramachandran, K. L. & Pepinsky, R. B. (1986) Nature (London) **320**, 77–81
8. De, B. K., Misoni, K. S., Lukas, T. J., Mroczkowski, B. & Cohen, S. (1986) J. Biol. Chem. **261**, 13784–13792
9. Pepinsky, R. B. & Sinclair, L. K. (1986) Nature (London) **321**, 81–84
10. Haigler, H. T., Schlaepfer, D. D. & Burgess, W. H. (1987) J. Biol. Chem. **262**, 6921–6930
11. Valentine-Braun, K. A., Hollenberg, M. D., Fraser, E. & Northup, J. K. (1987) Arch. Biochem. Biophys. **259**, 262–282
12. Tait, J. F., Gibson, D. & Fujikawa, K. (1989) J. Biol. Chem. **264**, 7944–7949
13. Sheets, E. E., Giugni, T. D., Coates, G. G., Schlaepfer, D. D. & Haigler, H. T. (1987) Biochemistry **26**, 1164–1172
14. Sawyer, S. T. & Cohen, S. (1985) J. Biol. Chem. **260**, 8233–8236
15. Giugni, T. D., James, L. C. & Haigler, H. T. (1985) J. Biol. Chem. **260**, 15081–15090
16. Cohen, S. & Fava, R. A. (1985) J. Biol. Chem. **260**, 12351–12358
17. Giugni, T. (1985) Ph.D. Thesis, University of California, Irvine.
18. McKanna, J. A. & Cohen, S. (1989) Science (Wash. D.C.) **243**, 1477–1479
19. Summers, T. A. & Creutz, C. E. (1985) J. Biol. Chem. **260**, 2437–2443
20. Khanna, N. C., Tokuda, M. & Waisman, D. M. (1986) Biochem. Biophys. Res. Commun. **141**, 547–554
21. Schlaepfer, D. D. & Haigler, H. T. (1988) Biochemistry **27**, 4253–4258
22. Michener, M. L., Dawson, W. B. & Creutz, C. E. (1986) J. Biol. Chem. **261**, 6548–6555
23. William, F., Haigler, H. T. & Kraft, A. S. (1989) Biochem. Biophys. Res. Commun. **160**, 474–479
24. Schlaepfer, D. D., Jones, J. & Haigler, H. T. (1992) Biochemistry **31**, 1886–1891
25. Schlaepfer, D. D. & Haigler, H. T. (1990) J. Cell Biol. **111**, 229–238
26. Ando, Y., Imamura, S., Hong, Y.-M., Owada, M. K., Kakunaga, T. & Kannagi, R. (1989) J. Biol. Chem. **264**, 6948–6955
27. Walter, P. & Lingappa, V. R. (1986) Annu. Rev. Cell Biol. **2**, 499–516
28. Auron, P. E., Webb, A. C., Rosenwasser, L. J., Mucci, S. V., Rich, A., Wolff, S. M. & Dinarello, C. A. (1984) Proc. Natl. Acad. Sci. U.S.A. **81**, 7907–7911
29. March, C. J., Mosley, B., Larsen, A., Cerretti, D. P., Braedt, G., Price, V., Gillis, S., Henney, C. S., Kronheim, S. R., Grabstein, K., Conlon, P. J., Hopp, T. P. & Cosman, D. (1985) Nature (London) **315**, 641–647
30. Abraham, J. A., Mergia, A., Whang, J. L., Tumolo, A., Friedman, J., Hjerrild, K. A., Gospodarowicz, D. & Fiddes, J. C. (1986) Science **233**, 545–548
31. Jaye, M., Howk, R., Burgess, W., Ricca, G. A., Chiu, I. M., Ravera, M. W., O'Brien, S. J., Modi, W. S., Maciag, T. & Drohan, W. N. (1986) Science **233**, 541–545
32. Muesch, A., Hartmann, E., Rohde, K., Rubartelli, A., Sitia, R. & Rapoport, T. A. (1990) Trends Biochem. Sci. **15**, 86–87
33. Moore, B., Zhang, Y., Adesnik, M. & Sabatini, D. D. (1988) J. Cell Biol. **107**, 699a
34. Singer, I. I., Scott, S., Hall, G. I., Limjuco, G., Chin, J. & Schmidt, J. A. (1988) J. Exp. Med. **167**, 389–407

35. Hazuda, D. J., Lee, J. C. & Young, P. R. (1988) J. Biol. Chem. **263**, 8473–8479
36. Rubartelli, A., Cozzolino, F., Talio, M. & Sitia, R. (1990) EMBO J. **9**, 1503–1510
37. Geisow, M. J., Childs, J., Dash, B., Harris, A., Panayotou, G., Sudhof, T. & Walker, J. H. (1984) EMBO J. **3**, 2969–2974
38. Glenney, J. R., Tack, B. & Powell, M. A. (1987) J. Cell Biol. **104**, 503–511
39. Zokas, L. & Glenney, J. R. (1987) J. Cell Biol. **105**, 2111–2121
40. Nakata, T., Sobue, K. & Hirokawa, N. (1990) J. Cell Biol. **110**, 13–25
41. Blackwood, R. A. & Ernst, J. D. (1990) Biochem. J. **266**, 195–200
42. Glenney, J. (1986) J. Biol. Chem. **261**, 7247–7252
43. Biemann, K. & Scoble, H. A. (1987) Science **237**, 992–998
44. Bradshaw, R. A. (1989) Trends Biochem. Sci. **14**, 276–279
45. Frey, B. M., Frey, F. J., Lingappa, V. R. & Trachsel, H. (1991) Biochem. J. **275**, 219–225
46. Pepinsky, R. B., Sinclair, L. K., Browning, J. L., Mattaliano, R. J., Smart, J. E., Chow, E. P., Falbel, T., Ribolini, A., Garwin, J. L. & Wallner, B. P. (1986) J. Biol. Chem. **261**, 4239–4246
47. Pepinsky, R. B., Tizard, R., Mattaliano, R. J., Sinclair, L. K., Miller, G. T., Browning, J. L., Chow, E. P., Burne, C., Huang, K. S., Pratt, D., Wachter, L., Hession, C., Frey, A. Z. & Wallner, B. P. (1988) J. Biol. Chem. **263**, 10799–10811
48. Ambrose, M. P. & Hunninghake, G. W. (1990) J. Appl. Physiol. **68**, 1668–1671
49. Smith, S. F., Tetley, T. D., Guz, A. & Flower, R. J. (1990) Environ. Health Perspect. **85**, 135–144
50. Flaherty, M. J., West, S., Heimark, R. L., Fujikawa, K. & Tait, J. F. (1990) J. Lab. Clin. Med. **115**, 174–179
51. Hullin, F., Raynal, P., Ragab-Thomas, J. M. F., Fauvel, J. & Chap, H. (1989) J. Biol. Chem. **264**, 3506–3513
52. Isacke, C. M., Lindberg, R. A. & Hunter, T. (1989) Mol. Cell. Biol. **9**, 232–240
53. Northup, J. K., Valentine-Braun, K. A., Johnson, L. K., Severson, D. L. & Hollenberg, M. D. (1988) J. Clin. Invest. **82**, 1347–1352
54. Christmas, P., Callaway, J., Fallon, J., Jones, J. & Haigler, H. T. (1991) J. Biol. Chem. **266**, 2499–2507
55. Fisher, E. R. & Sieracki, J. C. (1970) Pathol. Annu. **5**, 1–26
56. Srigley, J. R. & Hartwick, W. J. (1988) Ultrastruct. Pathol. **12**, 49–65
57. Stone, M. P., Stone, K. R., Ingram, P., Mickey, D. D. & Paulson, D. F. (1977) Urol. Res. **5**, 185–200
58. Vihko, P., Virkkunen, P., Henttu, P., Roiko, K., Solin, T. & Huhtala, M. J. (1988) FEBS Lett. **236**, 275–281
59. Mbikay, M., Nolet, S., Fournier, S., Benjannet, S., Chapdelaine, P., Paradis, G., Dube, J. Y., Tremblay, R., Lazure, C., Seidah, N. G. & Chretien, M. (1987) DNA **6**, 23–29
60. Lundwall, A. & Lilja, H. (1987) FEBS Lett. **214**, 317–322
61. Cooper, D. N. W. & Barondes, S. H. (1990) J. Cell Biol. **110**, 1681–1691
62. Hale, J. E. & Wuthier, R. E. (1987) J. Biol. Chem. **262**, 1916–1925
63. Genge, B. R., Wu, L. N. Y. & Wuthier, R. E. (1990) J. Biol. Chem. **265**, 4703–4710
64. Rothman, J. E. (1989) Cell **59**, 591–601

3

# Annexin I genes: diversity and regulation

**Nelson D. Horseman**

Department of Physiology and Biophysics, College of Medicine, University of Cincinnati, Cincinnati, OH 45267, U.S.A.

## Introduction

This paper will address two issues: the structure and diversity of *annexin I* (*anxI*) genes, and the regulation of *anxI* gene expression. I will describe some results regarding the regulation of annexin I in spite of the fact that we know very little regarding the function of the annexin I proteins and therefore cannot understand the physiological relevance of whatever 'regulation' we observe for the genes. Annexin I (AnxI) has been independently discovered at least five times [1–5] and in each case it has been possible to propose (however prematurely) some particular AnxI function. Understanding the regulation of *anxI* genes might provide some clues as to the proteins' functions if we assume that *anxI* regulation is an essential component of the physiological changes that take place during its regulation. However, such associations will always be correlative and have been the sources of many errant paths.

The first discovery of AnxI was the finding of Fava & Cohen [3] that the major cellular protein phosphorylated in A431 cells by the epidermal growth factor (EGF) receptor's protein-tyrosine kinase activity, was a polypeptide that they named 'p35' (see chapter 2 by Haigler & Schlaepfer). This observation, along with the fact that a 36000 $M_r$ protein (later identified as AnxII) is a major substrate of the Src protein-tyrosine kinase (see chapter 5 by Gerke) [6] led to early speculation that AnxI and II might be important for mediating the positive control of growth. Recent findings clearly indicate that AnxI proteins are highly expressed in differentiated, non-dividing cells in many instances, and that they may play roles in both growth and terminal differentiation. Fava and coworkers [7] examined a variety of rat tissues by immunohistochemistry and showed that mammalian AnxI was present in many terminally differentiated cell types and was not generally associated with dividing cells. On the other hand, Haigler and Schlaepfer [8] examined variations in three different Anx proteins during changes in cell growth state *in vitro* and observed that AnxI increased in cells induced to divide rapidly.

In the pigeon, $anxI_{cp35}$ is a unique gene that is stimulated by prolactin in the cropsac epithelium [1]. Although prolactin stimulates cell division in the proliferating basal layer, as well as differentiation in the distal layers of epithelium, $anxI_{cp35}$ is present exclusively in the differentiating cells [9]. This system provides the most compelling evidence that AnxI proteins are more likely to function in cell differentiation than as mediators of cell growth.

## Does glucocorticoid regulate mammalian AnxI?

Although the human AnxI protein (p35) had been discovered as the major substrate of the EGF receptor protein-tyrosine kinase in 1984 [3], the first sequence data that defined both AnxI (lipocortin I) and, by extension, the whole family of annexins, were presented in two papers by Pepinsky and Wallner and coworkers in 1986 [5,10]. A protein of ~ 37 kDa was purified from peritoneal lavage of rats injected with pharmacological doses of the synthetic glucocorticoid dexamethasone. It was proposed that the protein isolated was an important mediator of the anti-inflammatory actions of the glucocorticoids. Sequences of several tryptic and CNBr peptides were determined [5] and used to generate oligodeoxynucleotide probes with which to screen a human U937 cell cDNA library [10].

The peptide deduced from the human lipocortin I (AnxI) cDNA sequence was approximately 90 % similar to the rat peptide sequences. On the basis of the cDNA sequence the human AnxI polypeptide has an inferred molecular mass of 38 711 Da. Subsequently several other annexins, as well as AnxI from several different species, were cloned and sequenced, demonstrating the conserved motifs that characterized the family.

The original inference that AnxI was a glucocorticoid-responsive gene, and the implication that it mediated the anti-inflammatory actions of dexamethasone, have remained controversial and unproven (see Chapter 4 by Russo-Marie). Some features of the original observations were inconsistent or contradictory. AnxI protein was purified from the acellular portion of peritoneal lavage collected 1 h after dexamethasone treatment [5]. This material proved to be a rich source of AnxI protein. On the other hand, mRNA levels for AnxI were measured in cells from peritoneal exudates collected 2 h after dexamethasone treatment [10]. Under these experimental conditions there was no AnxI protein induction in the peritoneal exudate cells, but *anxI* mRNA was six-fold higher after dexamethasone treatment, compared with untreated animals. Recently, the rapid induction of the *anxI* mRNA was re-examined [11] and not replicated.

Several papers have shown that *anxI* mRNA is not generally induced by glucocorticoids in cells or tissues. In an extensive study of both

**Fig. 1. Schematic diagram of the organization of annexin I genes.**

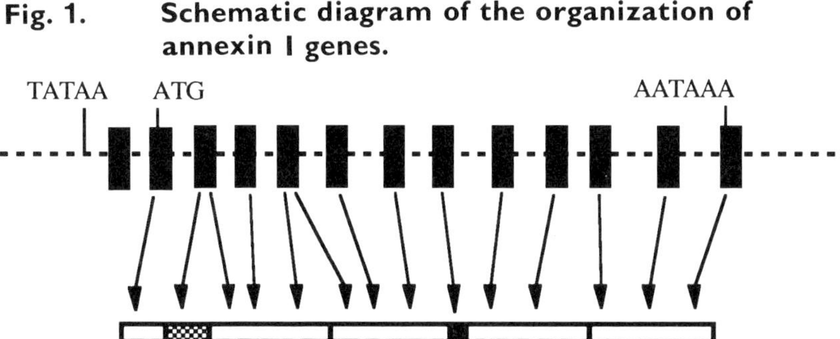


*The upper line depicts the 13-exon structure common to all of the known annexin I and annexin II genes. The exons are indicated by the solid black boxes and the introns by the dotted lines. The sizes of the exons and introns are not to scale. The TATA-box (promoter), initiator codon (encoded by ATG) and polyadenylation signal (AATAAA) are marked by labelled vertical lines. The lower line depicts the organization of the annexin I protein and its relationships with the exons of the gene. Arrows connect each exon with the region of the proteins encoded by the exon. N1 and N2 are the two subdomains of the N-terminal leader segment of the protein. Repeat units are indicated by roman numerals and the dark bar indicates the region connecting repeats 2 and 3 (encoded by exon 8).*

established human cell lines and primary cell cultures, Brönnegård *et al.* [12] demonstrated that 6 h after dexamethasone treatment *anxI* mRNA was not induced. Simultaneous measurement of metallothionine induction showed that the cells all contained functional glucocorticoid receptors and response systems. In astroglial cell cultures from rat brain, AnxI was not regulated by glucocorticoid treatment and glucocorticoid receptor antagonists did not suppress AnxI [13].

Recently, Wong *et al.* [11] studied the time course of glucocorticoid action on *anxI* mRNA and protein. In two established cell lines, but interestingly not in primary peritoneal macrophages, they demonstrated that glucocorticoid induced both *anxI* mRNA and its protein. Primary macrophages from peritoneal exudates were assayed after isolation from animals either treated with dexamethasone for 2 h or untreated, and the *anxI* mRNA level was not affected by dexamethasone treatment. In addition, treatment of adherent macrophages *in vitro* with dexamethasone (20 h) did not induce any increase of *anxI* mRNA. Because the experiments of Wallner *et al.* [10] and Wong *et al.* [11] used different species (rat and mouse, respectively) it is at least conceivable that the difference between them is a function of species specificity. If so, the mechanisms involved are unlikely to be of any general importance to the anti-inflammatory actions of glucocorticoids.

Although dexamethasone did not have any effect on *anxI* in primary macrophages, Wong *et al.* [11] did show that it increased the level of both protein and mRNA in 3T3-L1 cells and LA-4 lung epithelial cells. The induction by dexamethasone (1 μM) of *anxI* mRNA in 3T3-L1 cells began after a 2 h lag and continued through 24 h. By 24 h *anxI* mRNA had increased by about fivefold compared with basal levels. The increase of AnxI protein was delayed by several hours. The mRNA induction was blocked by simultaneous addition of cycloheximide. On the other hand, induction of metallothionein mRNA was not blocked by cycloheximide. If cycloheximide treatment was delayed until 4 h after the application of dexamethasone, it failed to block the increase of *anxI* mRNA. Therefore, protein synthesis is necessary within the first 4 h after glucocorticoid treatment in order for *anxI* mRNA to be stimulated by the hormone. These studies indicate that, in at least some cell lines, glucocorticoids can increase the expression of *anxI* mRNA. The mechanisms responsible for this effect are apparently secondary to genes that are induced during the initial phase of glucocorticoid action.

## Annexin I gene structures

The exon–intron organization of *anxI* genes from four species (one avian, three mammalian) have been described recently [14–16]. The genes each consist of 13 exons of which the first and last are uncoding 5′ and 3′ sequence, respectively (Fig. 1). The translation initiator codon (AUG) is encoded near the beginning of the second exon of each of the genes. Table 1 shows the exon sizes for each of the four *anxI* genes, and the human *anxII* gene [17].

The conservation of the intron–exon organization of the *anx* genes so far analysed is remarkable. The exons numbered 4–12 that span from the first repeat domain through the fourth repeat are identical in size amongst all of the four *anxI* genes. Compared with *anxII* (both mouse and human) this organization is also identical except that exon 8 includes one additional codon (three base pairs) in *anxII*. Exon 8 encodes amino acids that lie in the linker between the first and second halves of the Anx

**Table 1. Exon size comparisons of four annexin I genes and the human annexin II gene.**

| Exon | Pigeon *anxI* | Mouse *anxI* | Rat *anxI* | Human *anxI* | Human *anxII* |
|---|---|---|---|---|---|
| 1 | 47 | 61 | 61 | 61 | 44 |
| 2 | 79 | 80 | 79 | 79 | 59 |
| *3 | 94 | 109 | 109 | 109 | 100 |
| 4 | 95 | 95 | 95 | 95 | 95 |
| 5 | 114 | 114 | 114 | 114 | 114 |
| *6 | 91 | 91 | 91 | 91 | 91 |
| 7 | 80 | 80 | 80 | 80 | 80 |
| 8 | 57 | 57 | 57 | 57 | 60 |
| *9 | 94 | 94 | 94 | 94 | 94 |
| *10 | 96 | 96 | 96 | 96 | 96 |
| 11 | 59 | 59 | 59 | 59 | 59 |
| 12 | 123 | 123 | 123 | 123 | 123 |
| 13 | 235-polyA | | | | |

*The thirteenth exon includes the 3′ untranslated region and polyadenylation signal and its size is only tabulated for the pigeon* anxI *gene. *, Exons with a split codon at their 3′ end.*

proteins. All of the significant differences in organization amongst the *anxI* and *II* genes are within exons that encode the 5′ and 3′ untranslated regions (exons 1 and 13) and the *N*-terminal domains (exons 2 and 3).

Exon 2 is homologous between the pigeon and the three mammalian *anxI* genes, but differs in the *anxII* gene. The only exon 2 size difference amongst the four *anxI* genes is the murine gene in which the initiator codon is 15 rather than 14 bases into exon 2, resulting in one extra base in exon 2. The second exon is completely non-homologous between the *anxI* genes and *anxII*. It encodes a shorter region of the protein in *anxII* and has no sequence similarity with *anxI*.

Exon 3 is partially homologous amongst the *anxI* and *anxII* genes. The second portion of exon 3 encodes the *N*-terminal end of the conserved core domain. This portion was derived from a common ancestor in all of the *anxI* and *II* genes. When one compares the polypeptide sequences of all of the remaining annexins with the gene structures of *anxI* and *II*, it seems likely that the distal portion of exon 3 is conserved throughout the family. In all of the *anxI* and *II* genes, the distal end of exon 3 divides a codon. This codon, shared with exon 4, encodes one of the absolutely conserved glycine residues of the first 'Geisow' sequence. In the AnxV protein structure deduced by Huber *et al.* [18,19] the conserved glycines are present in the loop that forms an essential part of a high-affinity calcium-binding site. This suggests that any rearrangement of the gene which disturbed the split codon at the end of exon 3 would probably produce a non-functional protein, and be selected against.

The proximal (5′) end of exon 3 encodes the region that has least sequence similarity between the mammalian and pigeon *anxI* genes. Similarly, these portions of the *anxI* and *anxII* genes are the most different even though the distal portion of the third exon encodes part of the conserved core domain. It will be interesting to examine the apparent lability of the first half of exon 3 more broadly in other annexin genes and to determine how this variable region is related to the conserved core domain in several of the annexins.

The total size of the pigeon *cp35* gene that we cloned is about 15 kb. On the other hand, the mammalian *anxI* genes are longer (17–19 kb). The difference between the size of pigeon $anxI_{cp35}$ and the mammalian *anxI* genes lies in the fact that the first intron (i.e. between exons 1 and 2) of the mammalian genes is very long (4–6 kb) whereas intron I of *cp35* is about 1.5 kb. The available sequence information indicates that exon 1 and intron I are not homologous between the pigeon and mammalian genes. Therefore, if during the duplication and divergence that produced the pigeon $anxI_{cp35}$ gene the 5′ end was independently derived, regulation by prolactin may have been conferred by the unique 5′ sequences.

## Does the *anxI* gene organization correlate with protein structure?

Because the annexins are characterized by amino acid sequences that contain a fourfold symmetry [20,21] it has been speculated that the core structure of the annexins arose as a series of two duplications of an ancestral unit equivalent to one of the repeat units [22]. This notion has a natural attraction because of the analogy with the calmodulin family of proteins in which a small (12 amino acid) unit that forms the calcium-binding site is repeated four times. These ‘E-F hand’ units are believed to have arisen by two duplications of an ancestral calcium-binding domain [23]. The theorized order of those duplications can be inferred from the structure because domains 1 and 3, and 2 and 4, respectively, are nearest relatives of one another. In the simple case of the E-F hand proteins a duplication of an ancestral polypeptide gene apparently gave rise to a two-repeat unit, and that two-repeat unit subsequently duplicated to produce the current fourfold symmetrical molecule. Accordingly, the fourfold subdomain structure of the annexins has been hypothesized to have arisen by a similar series of steps. This notion is seductively simple and may even be right. However, the available data can be interpreted in many ways, and they do not convincingly eliminate any of several reasonable scenarios for *anx* gene evolution.

The structure of the annexins is not as simple and obvious as that of the calmodulin family of proteins. Firstly, the repeat unit is about 70 amino acids in length rather than 12–15. Secondly, the three-dimensional structure of AnxV indicates that the calcium-binding sites of the annexins are determined by discontinuous sequences consisting of several residues that are separated by about 40 amino acids [18,19]. Finally, and most importantly, the sequences of the four repeats do not fall into two sets of 'nearest relatives' which would allow one to infer the order of branching between them. At the polypeptide level the third repeat subdomain is least similar to any of the other repeats. Repeats 1, 2 and 4 are equally similar to each other and equally different from repeat 3 [24]. In all of the genes analysed so far each of the four annexin subdomains is encoded by sequences in three exons. It has been proposed that exon 5 was derived by fusion from exons that were ancestral to the current exons 10 and 11 [16]. It is equally, maybe more, plausible that a single ancestral exon was split by the insertion of an intron to yield exons 10 and 11.

We compared the nucleic acid sequnces of the exons in *cp35* that encode the highly conserved 'Geisow motifs' by a window-stringency method. Exons 4 and 6 (encoding the first and second repeated sequences) are most similar at the nucleotide sequence level. Exons 6 and 12 (repeats 2 and 4) and 4 and 12 (1 and 4) are somewhat less similar than 4 and 6. Exon 9 (third Geisow motif), while being most unlike any of the other exons, is much more similar to exon 6, than it is to either 4 or 12. These relationships suggest that the order of branching for the repeats is not consistent with any models in which repeats 1 and 3 (i.e. exons 4 and 9) must be closely related, such as that proposed by Kovacic and coworkers [16]. In addition, the sequence comparisons suggest either that selection has favoured a significant divergence of repeat 3, or that functional constraints have demanded preferential conservation of repeats 1, 2 and 4. Interestingly, high-affinity calcium-binding sites in anxV, mapped by X-ray crystallography, are present only in repeats 1, 2 and 4 [18,19].

The Anx protein structures, inferred from X-ray crystallography [18,19] or analysis of linear sequences [24], consist of a set of 20 highly ordered α-helical elements. Because of the functional constraints of this highly organized protein structure, it is possible that some of the apparent homology within the proteins has arisen by convergence, rather than having been derived from common ancestral sequences. This situation confronts the classic problem in making evolutionary inferences at any level, from organs to molecules.

Detailed sequence analysis at the nucleic acid level will be required to make confident inferences about the evolutionary history of the annexin-repeat structure. Until such an analysis is carried out we can only conclude that the genes encode proteins that have four related domains. Those domains share common structural features and have many similar, but some different, functions. At least parts of the $\sim$ 70-amino-acid repeated subdomain were probably derived from common ancestral sequences. Detailed models regarding the supposed evolutionary history of the repeat structures [16] are clearly speculative and premature, particularly if such models do not generate testable hypotheses.

## Special features of the pigeon $anxI_{cp35}$ gene

The pigeon $anxI_{cp35}$ gene is distinctive because it encodes an AnxI protein that lacks phosphorylation sites in the *N*-terminal leader. Because of this observation, the *N*-terminal leader of AnxI can be divided into two subdomains. The first is encoded by exon 2 and includes sequences that are shared between the pigeon cp35 and other AnxI proteins. This region can be thought of as diagnostic for AnxI proteins. The second portion of the *N*-terminal leader is encoded by the first half of exon 3 and is non-homologous between the pigeon and the mammalian genes. This second half is referred to as the 'hinge' because of its connection with the conserved *C*-terminal core region. The hinge can be hypothesized to play an important role in the regulation of Anx protein function. In AnxI and AnxII, it contains sites for phosphorylation [25]. These include protein kinase C sites and protein-tyrosine kinase sites that are at the beginning of the hinge (amino acid 21 of AnxI or 23 of AnxII) and are encoded by sequences in exon 3 of *anxII* and exon 2 of *anxI*.

The fact that the *N*-terminus can be divided into two segments that differ in their conservation amongst the AnxI proteins suggests that these two subdomains may have individual functional activities. In the AnxII protein, the first half of the *N*-terminal domain binds to the p11 small subunit of calpactin I [26] and the hinge region contains phosphorylation sites for regulation of the protein. If, by analogy, we consider that AnxI might be similarly organized, it is possible that the conserved first 20 amino acids of mammalian AnxI and pigeon cp35 might have a function such as binding to another polypeptide to form a multisubunit protein. This prediction is testable by several experiments, but so far there are no obvious candidates for subunits that bind to the AnxI *N*-terminus. It will be very informative to perform 'domain swapping' experiments that alter the *N*-terminal regions of specific annexins and to test the effects of different structures on recombinant protein function.

Unlike mammals, the pigeon genome encodes multiple *anxI* genes. By Southern hybridization, multiple genomic fragments hybridize with $anxI_{cp35}$ cDNA [15]. In addition to *cp35*, we have identified, by protein purification, cDNA cloning and genomic DNA cloning, a

**Fig. 2. Aligned sequences of three annexin I peptide sequences in the 'hinge' region (subdomain 2 of the *N*-terminus).**

```
Pigeon cp35   FMENLEQECIKCTQCVHGVPQQT.....NFDPSADVVALEK
    64%       |||  ||| ||       ||||      |||||| ||||||
Pigeon cp37   FMEHQEQEYIKSVKGGPVVPQQQ....PNFDPSAAVVALEK
    39%       | |  ||||   ||     |       | | ||  | || |
Human AnxI    FIENEEQEYVQTVKSSKGGPGSAVSPYPTFNPSSDVAALHK
```

*Identical amino acids are indicated by vertical bars and gaps are filled with periods. The percent of identical amino acids between each pair is also noted.*

second pigeon *anxI* gene ($anxI_{cp37}$) that encodes a protein that is phosphorylated by both tyrosine-protein kinase and protein kinase C (H. T. Haigler, J. Mangili, Y. Gao & N. D. Horseman, unpublished work). This protein has a sequence in the *N*-terminal domain which clearly indicates that it is a sibling of cp35 (Fig. 2). However, *cp37* has several specific features that make it more similar to mammalian *anxI*. The gene encodes a protein with substrate sites for the EGF receptor protein-tyrosine kinase and protein kinase C, and a trypsin-hypersensitive lysine within the hinge region of the *N*-terminus. Because the sequence of cp37 is homologous with cp35 but not mammalian AnxI, in the relevant hinge region, the property of protein kinase regulation of the AnxI *N*-terminus has evolved independently by convergence from different ancestral sequences. This is analogous to the independent evolution of protein kinase regulation by mammalian AnxI and AnxII.

Although the prolactin-dependent cp35 protein is not regulated by phosphorylation, the property of kinase regulation is apparently not completely dispensable as it is retained in the columbid cp37 polypeptide. It is, as yet, unknown whether multiple *anxI* genes are present in other avians, or are a unique feature of the family *Columbidae* (pigeons and doves). We are currently addressing this issue by cloning *anxI* genes from the chicken and other vertebrates. In humans there are multiple loci that contain *anxII* sequences [27]. However, these multiple loci include only a single functional gene and several pseudogenes (retroposons) [17].

## Promoter sequences and transcriptional regulation of *anxI* genes

For each of the *anxI* genes studied to date, 5′ flanking sequences of at least 450 bases have been published [14–16]. All of the genes contain a canonical 'TATA' promoter located approximately 30 bases upstream of their transcription start sites. The three mammalian sequences (human and two rodents) share significant similarities in the 5′ flanking region. These similarities are most apparent within 100 base pairs upstream of the transcription start site. As discussed above, the mouse *anxI* gene appears to be regulated indirectly by glucocorticoids [11]. This regulation has been suggested to depend upon a unique regulatory element that is completely different from the well characterized glucocorticoid response element (GRE) that mediates the rapid, direct response to activated glucocorticoid receptor. This alternative glucocorticoid-sensitive sequence has been referred to as the slow glucocorticoid response relement (sGRE) [11]. Horlick *et al.* [14] showed that the mouse *anxI* promoter region contains a sequence essentially identical to the sGRE just upstream of the promoter at bases −82 to −42. Both the human and rat *anxI* promoter regions published by Kovacic *et al.* [16] include a sGRE in identical locations in their most highly conserved 5′ flanking regions. The conservation of both sequence similarity and location argues that these sGREs are actually homologous and that they are likely to be functional in all three mammals. Experiments to test activity of the conserved sGRE have not yet been reported.

Various putative 'fast' GRE sequences (i.e. directly responsive to glucocorticoid receptor) have also been noted in each of the mammalian *anxI* gene structures. In the mouse gene, sequences similar to a consensus GRE were identified near −100 and −300 bases relative to the transcription start point [14]. A putative GRE was identified near +180 within intron I of the rat gene, and near +250 in intron I of the human gene [16]. Although each of these sequences have similarity to defined GREs, the rules for defining such similarity are always ambiguous and subjective. Therefore, it is impossible to assess the significance of any of these sequences without functional experiments. It is especially disquieting that the rat and mouse, closely related rodents in the family *Muridae*, appear to contain GREs that are not similar to one another. Because all of the proposed GRE sequences are non-homologous (being different in both sequence and location within the genes) it seems unlikely that any of those sequences define loci for meaningful regulation.

Prolactin specifically regulates the transcription of the $anxI_{cp35}$ gene [1]. We have addressed the function of the cp35 promoter through binding and transcription experiments. In one set of experiments we measured binding of proteins to the promoter region of the *cp35* gene by gel-mobility shifting and defined a promoter-proximal region upstream of the TATA box that binds several prolactin-induced nuclear factors from pigeon cropsac [28]. Transcription assays using the *cp35* promoter suggest that this region mediates a prolactin-dependent stimulation of *cp35* transcription. The localization of prolactin-responsive sequences in the $anxI_{cp35}$ promoter provides the first proven substrate for analysing transcriptional regulation by prolactin, and the first model of transcriptional regulation of an annexin gene in any specific physiological contest. The relevant transcription factors for the *cp35* gene are being analysed by purification and cloning.

## Perspective

The specific transcriptional regulation of the $anxI_{cp35}$ gene by prolactin associates annexin I with a highly differentiated cell function. The occurrence of both multiple and unique forms of AnxI in the pigeon provides a glimpse at the possible mechanisms which have, throughout evolution, promoted the growth of the annexin family of proteins. One can consider that multiple *anxI* genes in the pigeon are analogous with early stages in the divergence of other annexin genes. Clearly, properties of both conserved and divergent functions are expressed from the *anxI* genes of the pigeon. The fact that phosphorylation sites in the *N*-terminal hinge region have evolved independently in mammalian AnxI and pigeon cp37 serves as a useful model for considering the evolution of AnxI and AnxII.

Without the duplication of *anxI* to produce multiple genes in the pigeons and doves, it would have been impossible for prolactin to co-opt the pre-existing *anxI* gene without affecting the primitive function(s) of AnxI. Similarly, other opportunities for annexin genes to become associated with specific physiological activities must have depended on the multiplication of annexins within the superfamily. The divergence of structure and function of those genes has produced a clan of related annexins, each of which is about 45–50 % similar. As more is learned about the functions of the individual annexins it will be very interesting to discover the degree to which their activities overlap, or are distinctive. Their relative similarity at the level of gene and protein structure suggests that the annexins might have similar functions. However, their highly cell-specific expression argues for specialized, rather than similar, functions. The $anxI_{cp35}$ gene represents the highest degree of specialization for any annexin studied to date. Whether it represents a precedent, or an exception, will be learned by studying the functions and expression of other annexins in the context of specific cellular regulation.

## References


1. Horseman, N. D. (1989) Mol. Endocrinol. **3**, 773–779
2. Glenney, J. R., Jr., Tack, B. & Powell, M. A. (1987) J. Cell Biol. **104**, 503–511
3. Fava, R. A. & Cohen, S. (1984) J. Biol. Chem. **259**, 2636–2645
4. Giugni, T. D., James, L. C. & Haigler, H. T. (1985) J. Biol. Chem. **260**, 15081–15090
5. Pepinsky, R. B., Sinclair, L. K., Browning, J. L., Mattaliano, R. J., Smart, J. E., Chow, E. P., Falbel, T., Ribolini, A., Garwin, J. L. & Wallner, B. P. (1986) J. Biol. Chem. **261**, 4239–4246
6. Glenney, J. (1986) Proc. Natl. Acad. Sci. U.S.A. **83**, 4258–4262
7. Fava, R. A., McKanna, J. & Cohen, S. (1989) J. Cell. Physiol. **141**, 284–293
8. Haigler, H. T. & Schlaepfer, D. D. (1990) Prog. Clin. Biol. Res. **349**, 91–108
9. Horseman, N. D., Chen, X., Liu, L., Poyet, P. & Hitti, Y. (1992) Gen. Comp. Endocrinol. **85**, 405–414
10. Wallner, B. P., Mattaliano, R. J., Hession, C., Cate, R. L., Tizard, R., Sinclair, L. K., Foeller, C., Chow, E. P., Browning, J. L., Ramachandran, K. L. & Pepinsky, R. B. (1986) Nature (London) **320**, 77–81
11. Wong, W. T., Frost, S. C. & Nick, H. S. (1991) Biochem. J. **275**, 313–319
12. Brönnegård, M., Andersson, O., Edwall, D., Lund, J., Norstedt, G. & Carlstedt-Duke, J. (1988) Mol. Endocrinol. **2**, 732–739
13. Gebicke-Haerter, P. J., Schobert, A., Dieter, P., Honegger, P. & Hertting, G. (1991) J. Neurochem. **57**, 175–183
14. Horlick, K. R., Cheng, I. C., Wong, W. T., Wakeland, E. K. & Nick, H. S. (1991) Genomics **10**, 365–374
15. Hitti, Y. S. & Horseman, N. D. (1991) Gene **103**, 185–192
16. Kovacic, R. T., Tizard, R., Cate, R. L., Frey, A. Z. & Wallner, B. P. (1991) Biochemistry **30**, 9015–9021
17. Spano, F., Raugei, G., Palla, E., Colella, C. & Melli, M. (1990) Gene **95**, 243–251
18. Huber, R., Schneider, M., Mayr, I., Römisch, J. & Paques, E.-P. (1990) FEBS Lett. **275**, 15–21
19. Huber, R., Römisch, J. & Paques, E. (1990) EMBO J. **9**, 3867–3874
20. Geisow, M. J. (1986) FEBS Lett. **203**, 99–103
21. Pepinsky, R. B., Tizard, R., Mattaliano, R. J., Sinclair, L. K., Miller, G. T., Browning, J. L., Chow, E. P., Burne, C., Huang, K.-S., Pratt, D., Wachter, L., Hession, C., Frey, A. Z. & Wallner, B. P. (1988) J. Biol. Chem. **263**, 10799–10811
22. Crompton, M. R., Moss, S. E. & Crumpton, M. J. (1988) Cell **55**, 1–3
23. Klee, C. B., Crouch, T. H. & Richman, P. G. (1980) Annu. Rev. Biochem. **49**, 489–515
24. Barton, G. J., Newman, R. H., Freemont, P. S. & Crumpton, M. J. (1991) Eur. J. Biochem. **198**, 749–760
25. Brugge, J. S. (1986) Cell **46**, 149–150
26. Johnsson, N., Marriott, G. & Weber, K. (1988) EMBO J. **7**, 2435–2442
27. Huebner, K., Cannizzaro, L. A., Frey, A. Z., Hecht, B. K., Hecht, F., Croce, C. M. & Wallner, B. P. (1988) Oncogene Res. **2**, 299–310
28. Xu, Y.-H. & Horseman, N. D. (1992) Mol. Endocrinol. **6**, 375–383

# Annexins, phospholipase $A_2$ and the glucocorticoids

**Françoise Russo-Marie**

INSERM U 332, ICGM, 22 rue Méchain, 75014 Paris, France

## Introduction

Annexins belong to a new family of proteins discovered within the last decade by investigators from various fields of research as different as neurobiology, pharmacology and cell biology. These studies have led to the identification of a family of proteins with various putative functions. Indeed, chromobindins have been shown to be involved in exocytosis [1,2], calelectrins in synaptic vesicle traffic [3–5], and proteins I, II and III in cytoskeletal interactions [6,7]. In addition, p68 is known to act as a lymphocyte cytoskeletal protein [8–10], calcimedins as calcium-buffering proteins [11–12], calpactins as signalling molecules [13–36], vascular anticoagulants as anticoagulant proteins [37–39], and lipocortins as anti-phospholipase $A_2$ ($PLA_2$) and anti-inflammatory proteins [40–49]. Proteins of the same family have also been implicated in DNA replication [50] and in mineralization [51–53]. These proteins have been renamed 'annexins' followed by a number [54]. At present, eight annexins have been described in the mammalian species [55–57]. All the proteins are formed by a unique motif that has been duplicated during evolution and that is found four times in most annexins and eight times in one annexin (annexin VI). The *N*-terminal part is of various length (from four amino acids in annexin IV to 167 amino acids in annexin VII), and carries the phosphorylation sites. Annexins possess a certain number of properties that are common to all of them and seem to be the consequence of their ability to bind calcium and negatively charged phospholipids [58]. Among those properties, their ability to inhibit $PLA_2$ seems of major importance for the understanding of their anti-inflammatory functions and their putative role of 'messengers' of glucocorticoid effects. I shall address the question of their inhibitory effect on $PLA_2$ in four steps: (i) How do annexins inhibit $PLA_2$? (ii) Can annexins inhibit the $PLA_2$ involved in signal transduction? (iii) Annexins inhibit $PLA_2$ in *in vitro* assays; could some *in vivo* situations mimic the *in vitro* situation? (iv) What is the connection between annexins and glucocorticoids?

## How do annexins inhibit phospholipase $A_2$?

### Historical background

Lipocortin (defined at that time as a biological entity that probably corresponds to annexin I) was initially described as an inhibitor of phospholipase $A_2$ whose synthesis and release were under the control of glucocorticosteroids [40–46]. This property was the basis of the biological assay used to identify the protein. The assay used as a substrate, [$^3$H]oleic acid-labelled *Escherichia coli* membranes, and as an enzyme, pancreatic porcine $PLA_2$; it was performed in the presence of 1 mM calcium [44]. The activity of the protein was estimated as the percentage of inhibition of total hydrolysis induced by $PLA_2$ alone versus $PLA_2$ in the presence of the protein. Analysis of the data showed excellent and perfectly reproducible dose-response curves of inhibition of $PLA_2$. The most relevant hypothesis was that lipocortin was a specific inhibitor of $PLA_2$, although in most conditions, it was not possible to demonstrate a direct interaction between $PLA_2$ and lipocortin. These studies were carried out with no prior knowledge of the basic property of annexins to bind to negatively charged phospholipids in the presence of calcium. When the full sequence of lipocortin (annexin I) was determined, showing that it belonged to a family of proteins possessing such properties, a new hypothesis appeared concerning the characteristics of $PLA_2$ inhibition.

### Evidence that annexins do not inhibit $PLA_2$ through a direct protein–protein interaction but rather by interaction with phospholipids

When it was realized that the basic property of annexins is binding to phospholipids in the presence of calcium [59–61], it was hypothesized that they must interact directly with phospholipids rather than with $PLA_2$ itself. Some observations favoured this hypothesis; indeed, a direct interaction between annexins and $PLA_2$ has never been found in *in vitro* conditions, i.e. in a test tube (F. Russo-Marie, unpublished work) [62], but only after passage on columns coated with the enzyme or another control protein, suggesting that the annexins had probably interacted with the hydrophobic substrate rather than with the enzyme. It has since been clearly demonstrated that annexins were interacting with the substrate, as $PLA_2$ inhibition could be overcome by adding extra substrate, indicating that the enzyme was indeed still active, but that annexins were depleting $PLA_2$ of its substrate [63–65]. Additional experiments confirmed these reports. If zwitterionic phospholipids—these phospholipids do not bind annexins—are used as the substrate, $PLA_2$ is not inhibited [66], and a $PLA_2$ that does not require calcium for hydrolysing phospholipids is not inhibited by annexins [67]. This indicates that $PLA_2$ inhibition is the consequence of the binding of annexins to phospholipids in the presence of calcium.

Two requirements seem necessary for inhibition to occur: (i) phospholipids have to be negatively charged and (ii) calcium has to be present. This suggests that some specificity could be brought about by either the species of phospholipid or by the calcium concentration necessary for different annexins to bind to phospholipids.

## Arguments that still challenge the hypothesis of a direct interaction between $PLA_2$ and annexins

Although experiments reported in chapter 3 by Horseman are clearly straightforward in demonstrating that annexins do indeed bind to negatively charged phospholipids in the presence of calcium, and consequently inhibit $PLA_2$, other reports suggest that in addition to this mechanism, a direct effect of annexins on $PLA_2$ can still be envisaged. One report originated from the finding that lipocortin (or annexin I) and uteroglobin were both inducible by glucocorticoids and both behaved as inhibitors of $PLA_2$ [68]. A computer search for sequence similarity between these two proteins revealed a sequence of nine amino acids located within the third repeat of lipocortin and within the third α-helix of uteroglobin [69]. These peptides (HDMNKVLDL and MQMKKVLSD) were shown to be capable of inhibiting $PLA_2$ directly and were found to possess anti-inflammatory properties [69]. They have since been named antiflammins. However, contradictory reports have appeared that failed to observe the same data as the original papers [70–74]. Nevertheless, additional studies still challenge the concept. Indeed, another peptide (amino acids 204–212) designed from the same region of the third repeat of annexin V is able to inhibit contraction of isolated rat stomach strips elicited by $PLA_2$, to reduce prostaglandin E2 release from isolated cells and to possess anti-inflammatory properties [75], although the peptide does not directly inhibit the enzyme in the *in vitro* assay. Moreover, monoclonal antibodies raised against annexin I, and which react with the $Ca^{2+}$-bound form of this annexin, abolish the inhibition of $PLA_2$ induced by annexin I, although they do not affect its binding to *E. coli* membranes or to phosphatidylserine [76].

## Discussion

These experiments suggest that, in *in vitro* assays, annexins inhibit $PLA_2$ by interacting calcium-dependently with negatively charged phospholipids, the substrate of the enzyme, thus depriving the enzyme of its substrate.

$PLA_2$ that is acting on zwitterionic phospholipids is not inhibited by annexins [66]. These data are in strong agreement with experiments that have analysed the binding of annexins to various types of phospholipids in the presence of calcium. Inhibition of $PLA_2$ correlates with the ability of annexins to bind to different types of phospholipids. Nevertheless, no real difference was found between all annexins in relation to their abilities to inhibit $PLA_2$ [66,77], whereas differences have been found in their abilities to bind to different types of phospholipids [78–81]. This may be because of the concentration of calcium used for these various experiments which was in the millimolar range for analysing $PLA_2$ inhibition and in the micromolar range for the studies of binding to various phospholipids. In the millimolar range of calcium concentration, there is no difference in the binding of different annexins to any type of negatively charged phospholipids. This might explain why there is no difference in the assay used to measure their inhibitory effect on $PLA_2$.

According to the characteristic of annexins to bind to negatively charged phospholipids in the presence of calcium, one might expect them to inhibit other phospholipases and/or enzymes that require calcium-dependent binding to similar substrates. There have been no reports supporting this hypothesis although, in one report, annexins I and II were shown to inhibit phosphatidylinositol (PI) and PI-bisphosphate-specific phospholipase C, with either PI or PI-bisphosphate as substrates [82]. However, the weak interaction of annexins with PI at physiological calcium concentrations and the absence of interactions of annexins IV and VI with PI-bisphosphate suggest that annexins probably do not interact with phospholipase C in physiological conditions [81]. In addition, proteins related to annexins were found to inhibit protein kinase C, although the putative mechanism of action did not seem to involve phospholipid and/or calcium binding and the proteins were not well characterized [83]. In our experiments, purified annexins I, II, V and VI were not found to be capable of inhibiting protein kinase C activity (C. Comera, unpublished work). These data raise the possibility that there might be a certain specificity for $PLA_2$ that could be the result of complex interactions between annexins, negatively charged phospholipids and the enzyme, that bring some specificity to these interactions. The effect of antiflammins on $PLA_2$ activity has not been shown consistently, especially in *in vitro* assays, although controversy continues; indeed, the antiflammin peptide derived from annexin I (but not the uteroglobin peptide) was shown to inhibit $PLA_2$ using a lipid-monolayer system [74]. The effect of antiflammins as anti-inflammatory agents is thus very controversial. It is therefore difficult to propose any interpretation concerning their mechanism of action. In addition, the crystal structure of annexin I is not yet known and, although one might expect similarities with annexin V [84–86] (see chapter 10 by Huber *et al.*), the crystal structure was not analysed in the presence of phospholipids. Therefore, any conclusion drawn from the crystal structure may be irrelevant for interpreting a putative mechanism of action for inhibiting $PLA_2$.

## Do annexins inhibit the $PLA_2$ involved in signal transduction?

Although at present there is no clear answer concerning regulation of the $PLA_2$ involved in signal transduction, understanding the characteristic of annexins to bind to different subspecies of phospholipids in relation to the calcium concentration permits some speculation. Indeed, specificity of binding to various subspecies of phospholipids may be brought about by (i) the annexin involved, (ii) the species and/or form of phospholipids (either in micellar, liposomes or large unilamellar vesicle forms), and (iii) the concentration of calcium required for binding. These differences cannot be taken into account in *in vitro* assays for soluble $PLA_2$ that require calcium in the millimolar range and, in such conditions, inhibition seems rather unspecific. However, for cellular $PLA_2$ that is active at physiological calcium concentrations within the cytoplasm, a certain specificity could occur. Arguments for proposing that specificity may exist are as follows. The intracellular concentration of calcium in the cytoplasm is 0.01–0.1 μM in resting conditions; following stimulation of the cell, this concentration can rise to 1 μM and very rarely to 10 μM [87]. This rise is shortlived and the calcium concentration oscillates within the cytoplasm [88]. The increase is secondary to two types of events: either the opening of calcium channels from endoplasmic reticulum after inositol trisphosphate binding to its receptor, or the opening of calcium channels in the plasma membrane. In both cases, the calcium concentration may increase in very discrete places to higher levels (close to the channels); in such cases the real calcium concentration is not known but is very transient and most certainly does not spread all over the membrane. It is highly unlikely, therefore, that endogenous intracellular $PLA_2$ cleaves phospholipids in the presence of 1 mM calcium as demonstrated *in vitro*. Indeed, it seems that cellular $PLA_2$s [89–92] are different enzymes from the extracellular soluble ones [93], as far as their structure, substrate specificity and calcium requirements are concerned. In view of this, the intracellular situation may be envisioned as bringing some specificity to the four participants in the reaction—$PLA_2$, phospholipids, calcium and annexins. The hypothesis that annexins may play a role in regulating cellular $PLA_2$ could therefore be relevant. Data showing that through phosphorylation, annexins might regulate transduction pathways involving $PLA_2$ [94] may therefore be relevant to intracellular situations.

In summary, there are no persuasive arguments for implicating annexins as intracellular regulators of $PLA_2$ activity, but a possible role for some annexins as $PLA_2$ inhibitors still exists. Such a function may be analogous to the action of profilin on phospholipase-C$\gamma$ activity [95].

## Annexins inhibit $PLA_2$ in *in vitro* systems; could some *in vivo* situations mimic the *in vitro* situation?

Previously, I have reported a series of studies performed *in vitro*, allowing the analysis of some biochemical and biophysical properties of the annexins. The relative ease with which large amounts of annexins may be purified, either from natural or recombinant sources, has allowed their use in pharmacological experiments, as well as the analysis of their effects on cellular functions related to $PLA_2$ activity, and their application to *in vivo* experiments.

### Effects of annexins on cellular functions related to $PLA_2$ activity

A series of cellular events is related to $PLA_2$ activation in cells involved in the inflammatory reaction. These events are measurable and include arachidonate release from prelabelled cells, formation of prostaglandins, leukotrienes and paf-acether, cell chemotaxis and superoxide release. Indeed, pretreatment of cells with various concentrations of annexins inevitably leads to a dose-response inhibitory curve of all the measurable parameters related to $PLA_2$ activation. These effects occurred for concentrations of annexins in the range of $10^{-8}$ M – $10^{-5}$ M. They were reversed by treating the annexins with specific antibodies. Not all annexins have been tested in all of these assays and annexin I (lipocortin) has been used most often. In experiments using other annexins such as annexin V or III, however, no differences were found in the cellular responses [96–100].

### Effects of annexins in the whole animal

In various models, annexins have been used for controlling the inflammatory response. The rationale for using annexins as anti-inflammatory agents came from earlier experiments implicating them as messengers of anti-inflammatory glucocorticoids and as a consequence of their ability to inhibit cellular events mediated by $PLA_2$ activation. Indeed, in different models annexins either used intravenously or applied directly to the inflammatory site induced a dose-response inhibition of cell migration and of the production of pro-inflammatory mediators produced by $PLA_2$ activation [101–102].

## Discussion

Knowing that the annexins bind to negatively charged phospholipids in the presence of calcium, how does one interpret results such as the dose-dependent inhibition of cellular events secondary to $PLA_2$ activation. In these models, how are annexins inhibiting $PLA_2$ activation? Do they bind to membranes and, if so, how do they communicate information to the other face of the membrane, where phospholipid cleavage occurs under the control of cellular $PLA_2$? A few hypotheses can be proposed that could explain their effect. First, annexins bind to membranes and by an unknown mechanism transmit information to intracellular $PLA_2$ blocking its activation. Second, annexins bind to a specific receptor on the cell membrane that transduces a message that inhibits intracellular $PLA_2$ activation. Third, annexins enter the cells and bind to phospholipids preventing their hydrolysis by intracellular $PLA_2$. Fourth, annexins trap lipid-derived metabolites such as prostaglandins, leukotrienes or paf-acether, thus preventing their autocrine and paracrine action on cells [77,81]. Finally, annexins bind to extracellular phospholipids and prevent an extracellular $PLA_2$ from hydrolysing phospholipids. Although there is no proof that any of these mechanisms occurs, two of them seem to be more plausible. Firstly, receptors for annexin I may reside on the surface of neutrophils and monocytes, cells where annexins do indeed inhibit the production of $PLA_2$ activity-derived mediators (N. J. Goulding, personal communication). The second plausible mechanism is that during inflammation $PLA_2$ is secreted, and indeed secreted forms of $PLA_2$ are well documented [103–104]. If secreted forms of $PLA_2$ are present and therefore able to hydrolyse phospholipids from the cell surface, the obligatory corollary is that activated cells should expose at their surface negatively charged phospholipids. In these conditions, and in the presence of 1 mM calcium (the actual extracellular calcium concentration), annexins would bind to phospholipids and prevent their hydrolysis by extracellular $PLA_2$.

Although it is not easy to understand these effects of annexins, it is difficult to assign a role to them that does not involve binding to phospholipids and would not involve $PLA_2$.

## Annexins and glucocorticoids: historical background

Members of the annexin family were initially described as the putative messengers of glucocorticoid action. Many experiments had demonstrated that glucocorticoids were anti-inflammatory (in their early action) because they inhibited the production of $PLA_2$ activity-derived mediators [40–49]. These effects required glucocorticoid receptor occupancy as well as *de novo* protein synthesis. A protein was found that was secreted by cells after glucocorticoid treatment and that inhibited $PLA_2$ in an *in vitro* assay (as described in the first paragraph) [42–45]. This led to the purification and cloning of the first annexin (named annexin I) [47–48].

As mentioned in the previous sections, annexins do indeed inhibit $PLA_2$ but are not direct specific inhibitors of the enzyme, thus questioning the specificity of annexins as $PLA_2$ inhibitors. Moreover, it had been demonstrated that this glucocorticoid-induced $PLA_2$ inhibitory protein was found in the supernatant of cells, as well as in the peritoneal lavage fluid of glucocorticoid-treated rats, demonstrating that it was secreted. None of the annexins have a signal sequence to support their secretion (but see chapter 2 by Haigler & Schlaepfer). In addition, the regulation of their synthesis under glucocorticoid treatment has not yet been seen clearly; further studies are required. Nevertheless, at least one annexin (annexin I or lipocortin) has in its promoter region a consensus sequence for a glucocorticoid-response element and, under certain conditions, the mRNA of lipocortin is increased by glucocorticoids [48,106].

It is not known whether other annexins are under the genomic control of glucocorticoids.

Despite difficulties encountered in demonstrating that one annexin is an unequivocal mediator of the anti-inflammatory effect of glucocorticoids, a series of arguments do indeed favour a possible role of annexin I as the second messenger of the anti-inflammatory effects of glucocorticoids. Thus, (i) annexin I is synthesized and released in certain conditions by cells after glucocorticoid treatment, (ii) annexin I mimics anti-inflammatory effects of glucocorticoids in whole cell assays or when administered *in vivo*, and (iii) some effects of glucocorticoids have been shown to be reversed by anti-lipocortin antibodies [107]. Nevertheless, important questions remain to be answered concerning the mechanism of action of lipocortin: are these effects the consequence of $PLA_2$ inhibition or other mechanisms and how is the protein translocated to the outside of the cell?

## Annexin I may possess other autocrine functions when it is translocated from the cell onto the membrane

Recent experiments performed in rat brain have demonstrated that annexin I is present in certain glial cells and neurons [108–111]. The *N*-terminal fragment (amino acids 1–188) of annexin I injected *in vivo* into the rat brain inhibits the central and peripheral actions of cytokines on fever and thermogenesis [112–113]. Inhibition of the actions of endogenous annexin I by cerebral injections of neutralizing antibody raised to the annexin I fragment, significantly reverses the inhibitory effects of glucocorticoids on cytokine action [114]. Intracerebroventricular administration of annexin I causes marked inhibition of cerebral oedema and infarct size after cerebral ischaemia. Ischaemia caused increased expression of annexin I around the area of infarction, and intracerebroventricular injection of neutralizing anti-annexin I fragment antiserum increases the size of the infarct.

These data demonstrate quite clearly that important functions of annexin I may involve protecting cells by altering the response of those cells to various stimuli. At present, it is not known whether these effects implicate $PLA_2$ inhibition or whether they are the consequence of binding to extracellular, exposed negatively charge phospholipids, and also whether other annexins might exhibit the same protective effects.

## Conclusions

Annexins are fascinating proteins, whose putative functions are so diverse that one might think they are playing with our rationality. Indeed, they seem to change function according to their environment, i.e., the nature of both the phospholipid species and phospholipid configuration, and most certainly the calcium concentration. These parameters are highly variable during cell activation and are very different within the various sub-compartments of cells. Thus, we may postulate that the functions of annexins will vary with the calcium concentration and that, allowing the accomplishment of cellular functions that are the consequence of calcium waves. At high calcium concentration, as in the extracellular milieu, the functions of annexins are probably different to those exhibited intracellularly, as their privileged partners, calcium and phospholipids, are so different. Nevertheless, one important issue has to be addressed: how are these proteins translocated across membranes and under what circumstances?

### References

1. Creutz, C. E. (1981) Biochem. Biophys. Res. Commun. **103**, 1395–1399
2. Creutz, C. E., Drust, D. S., Hamman, H. C., Junker, M., Kambouris, M. G., Klein, G. R., Nelson, M. R. & Snyder, S. L. (1990) in Stimulus Response Coupling: The Role of Intracellular Calcium-binding Proteins (Smith, V. L. & Dedman, J. R., eds), pp. 279–310, CRC Press, Boca Raton
3. Walker, J. H. (1982) J. Neurochem. **39**, 815–819
4. Südhof, T. C., Ebbecke, M., Walker, J. H., Fritsche, U. & Boustead, C. (1984) Biochemistry **23**, 1103–1108
5. Boustead, C. M., Geisow, M. J., Taylor, W. R. & Walker, J. H. (1990) in Stimulus Response Coupling: The Role of Intracellular Calcium-binding Proteins (Smith, V. L. & Dedman, J. R., eds) pp 251–279, CRC Press, Boca Raton
6. Shadle, P. J., Gerke, V. & Weber, K. (1985) J. Biol. Chem. **260**, 16354–16359
7. Gerke, V., Johnsson, N. & Weber, K. (1990) in Stimulus Response Coupling: The Role of Intracellular Calcium-binding Proteins (Smith, V. L. & Dedman, J. R., eds), pp. 311–339, CRC Press, Boca Raton
8. Owens, J. O. & Crumpton, M. J. (1984) Biochem. J. **219**, 309–314
9. Owens, J. O., Gallagher, C. J. & Crumpton, M. J. (1984) EMBO J. **3**, 945–950

10. Moss, S. E., Edwards, H. C. & Crumpton, M. J. (1990) in Stimulus Response Coupling: The Role of Intracellular Calcium-binding Proteins (Smith, V. L. & Dedman, J. R., eds), pp. 501–521, CRC Press, Boca Raton
11. Moore, P. B. & Dedman, J. R. (1982) J. Biol. Chem. **257**, 9663–9668
12. Kaetzel, M. A., Hazarika, P. & Dedman, J. R. (1990) in Stimulus Response Coupling: The Role of Intracellular Calcium-binding Proteins (Smith, V. L. & Dedman, J. R., eds), pp. 383–411, CRC Press, Boca Raton
13. Radke, K. & Martin, G. S. (1979) Proc. Natl. Acad. Sci. U.S.A. **76**, 5212–5216
14. Radke, K., Gilmore, T. & Martin, G. S. (1980) Cell **21**, 821–828
15. Erikson, E. & Erikson, R. L. (1980) Cell **21**, 829–839
16. Cooper, J. A. & Hunter, T. (1981) Mol. Cell. Biol. **1**, 165–171
17. Cooper, J. A. & Hunter, T. (1981) Mol. Cell. Biol. **1**, 394–399
18. Cheng, Y. S. E. & Chen, L. B. (1981) Proc. Natl. Acad. Sci. U.S.A. **78**, 2388–2392
29. Decker, S. J. (1982) Biochem. Biophys. Res. Commun. **109**, 434–439
20. Nakamura, K. D. & Weber, M. J. (1982) Mol. Cell. Biol. **2**, 147–151
21. Greenberg, M. E., Edelman, G. M. (1983) J. Biol. Chem. **258**, 8497–8502
22. Cooper, J. A. & Hunter, T. (1983) Curr. Top. Microbiol. Immunol. **107**, 125–132
23. Cooper, J. A. & Hunter, T. (1983) J. Biol. Chem. **258**, 1108–1113
24. Gould, K. L., Cooper, J. A. & Hunter, T. (1984) J. Cell. Biol. **98**, 487–492
25. Greenberg, M. E., Brackenbury, R. & Edelman, G. M. (1984) J. Cell. Biol. **98**, 473
26. Courtneidge, S., Ralston, R., Alitalo, K. & Bishop, J. M. (1983) Mol. Cell. Biol. **3**, 340–347
27. Greenberg, M. E. & Edelman, G. M. (1983) Cell **33**, 767–772
28. Nigg, E. A., Cooper, J. A. & Hunter, T. (1983) J. Cell. Biol. **96**, 1601–1606
29. Lehto, V. P., Virtanen, I., Passivuo, R., Ralston, R. & Alitalo, K. (1983) EMBO J. **2**, 1701–1706
30. Radke, K., Carter, V. C., Moss, P., Dehazya, P., Schliwa, M., Martin, G. S. (1983) J. Cell. Biol. **97**, 1601–1607
31. Glenney, J. R., Jr. (1986) Proc. Natl. Acad. Sci. U.S.A. **83**, 4258–4262
32. Gerke, V. & Weber, K. (1984) EMBO J. **3**, 227–232
33. Erikson, E., Tomaciewicz, H. G. & Erikson, R. L. (1984) Mol. Cell. Biol. **4**, 77–82
34. Fava, R. A. & Cohen, S. (1984) J. Biol. Chem. **259**, 2636–2641
35. Sawyer, S. T. & Cohen, S. (1985) J. Biol. Chem. **260**, 8233–8237
36. Guigni, T. D., James, L. C. & Haigler, H. T. (1985) J. Biol. Chem. **260**, 15081–15086
37. Funakoshi, T., Heimark, R. L., Hendrikson, L. E., McMullen, B. A. & Fujikawa, K. (1987) Biochemistry **26**, 5572–5577
38. Funakoshi, T., Hendrikson, L. E., McMullen, B. & Fujikawa, K. (1987) Biochemistry **26**, 8087–8092
39. Grundmann, U., Abel, K-J., Bohn, H., Löbermann, H., Lottspeich, F. & Hüpper, H. (1988) Proc. Natl. Acad. Sci. U.S.A. **85**, 3708–3712
40. Flower, R. J. (1984) Adv. Inflammation Res. **8**, 1–16
41. Hirata, F., Schiffman, E., Venkatasubramanian, K., Salomon, D., Axelrod, J. A. (1980) Proc. Natl. Acad. Sci. U.S.A. **77**, 2533–2537
42. Blackwell, G. J., Carnuccio, R., DiRosa, M., Flower, R. J., Parente, L., Persico, P. (1980) Nature (London) **287**, 147–150
43. Cloix, J. F., Colard, O., Rothhut, B. & Russo-Marie, F. (1983) Br. J. Pharmacol. **79**, 313–318
44. Rothhut, B., Russo-Marie, F., Wood, J., DiRosa, M. & Flower, R. J. (1983) Biochem. Biophys. Res. Commun. **117**, 878–882
45. Blackwell, G. J., Carnuccio, R., DiRosa, M., Flower, R. J., Langham, C. S. J., Parente, L., Persico, P., Russell-Smith, N. C. & Stone, D. (1982) Br. J. Pharmacol. **76**, 18–23
46. DiRosa, M., Flower, R. J., Hirata, F., Parente, L. & Russo-Marie, F. (1984) Prostaglandins **28**, 441
47. Pepinsky, R. B., Sinclair, L. K., Browning, J. L., Mattaliano, R. J., Smart, J. E., Chow, E. P., Falbel, T., Ribolini, A., Garwin, J. L., Wallner, B. P. (1986) J. Biol. Chem. **261**, 4239–4243

48. Wallner, B. P., Mattaliano, R. J., Hession, C., Cate, R. L., Tizard, R., Sinclair, L. K., Foeller, C., Chow, E. P., Browning, J. L., Ramachandran, K. L. & Pepinsky, R. B. (1986) Nature (London) **320**, 77–81
49. Huang, K. S., Wallner, B. P., Mattaliano, R. J., Tizard, R., Burne, C., Frey, A., Hession, C., McGray, P., Sinclair, L. K., Chow, E. P., Browning, J. L., Ramachandran, K. L., Tang, J., Smart, J. E. & Pepinsky, R. B. (1986) Cell **46**, 191–197
50. Jindal, H. K., Chaney, W. G., Anderson, C. W., Davis, R. G. & Vishwanatha (1991) J. Biol. Chem. **266**, 5169–5176
51. Genge, B. R., Wu, L. N. Y. & Wuthier, R. E. (1989) J. Biol. Chem. **264**, 10917–10921
52. Genge, B. R., Wu, L. N. Y. & Wuthier, R. E. (1990) J. Biol. Chem. **264**, 4703–4710
53. Genge, B. R., Wu, L. N. Y., Adkisson, H. D. & Wuthier, R. E. (1991) J. Biol. Chem. **266**, 10678–10685
54. Crumpton, M. J. & Dedman, J. R. (1990) Nature (London) **345**, 212
55. Pepinsky, R. B., Tizard, R., Mattaliano, R. J., Sinclair, L. K., Miller, G. T., Browning, J. F., Chow, E. P., Burne, C., Huang, K. S., Pratt, D., Wachter, L., Hession, C., Frey, A. Z. & Wallner, B. P. (1988) J. Biol. Chem. **263**, 10799–10803
56. Burns, A. L., Magendzo, K., Shirvan, A., Srivastava, M., Rojas, E., Alijani, M. R. & Pollard, H. B. (1989) Proc. Natl. Acad. Sci. U.S.A. **86**, 3798–3802
57. Hauptmann, R., Maurer-Fogy, I., Krystek, E., Bodo, G., Andree, H. & Reutelingsperger, C. P. M. (1989) Eur. J. Biochem. **185**, 63–71
58. Russo-Marie, F. (1991) Prostaglandins Leukotrienes Essential Fatty Acids, **42**, 83–89
59. Glenney, J. R., Jr, Tack, B. & Powell, M. A. (1987) J. Cell Biol. **104**, 503–509
60. Soric, J. & Gordon, J. A. (1986) J. Biol. Chem. **261**, 14490–14495
61. Schlaeffer, D. S., Mehlman, T., Burgess, W. H. & Haigler, H. T. (1987) Proc. Natl. Acad. Sci. U.S.A. **84**, 6078–6083
62. Ahn, N. G., Teller, D. C., Bienkowski, M. J., McMullen, B. A., Lipkin, E. W. & de Haen, C. (1988) J. Biol. Chem. **263**, 18567–18663
63. Davidson, F. F., Dennis, E. A., Powell, M. & Glenney, J. R. (1987) J. Biol. Chem. **262**, 1698–1703
64. Aarsman, A. J., Mynbeek, G., van den Bosch, H., Rothhut, B., Prieur, B., Comera, C., Jordan, L. & Russo-Marie, F. (1987) FEBS Lett. **219**, 176–179
65. Haigler, H. T., Schlaepfer, D. D. & Burgess, W. H. (1987) J. Biol. Chem. **262**, 6921–6926
66. Comera, C., Rothhut, B. & Russo Marie, F. (1990) Eur. J. Biochem. **188**, 139–146
67. Gassama-Diagne, A., Fauvel, J. & Chap, H. (1990) J. Biol. Chem. **265**, 4309–4314
68. Levin, S. W., Butler, J. D., Schumacher, U. K., Wightman, P. D. & Mukherjee, A. B. (1986) Life Sci. **38**, 1813–1819
69. Miele, L., Cordella-Miele, E., Facchiano, A. & Mukherjee, A. B. (1988) Nature (London) **335**, 726–730
70. Van Binsbergen, J., Slotboom, A. J., Aarsman, A. J. & De Haas, G. H. (1989) FEBS Lett. **247**, 293–297
71. Marki, F., Pfeilschifter, J., Rink, H. & Wiesenberg, I. (1990) FEBS Lett. **264**, 171–175
72. Hope, W. C., Patel, B. J. & Bolin, D. R. (1990) FASEB J. **4**, A2230
73. Travé, G., Lees, D., Liautard, J. P. & Widada, J. S. (1991) FEBS Lett. **293**, 34–36
74. Barton, G. J., Newman, R. H., Freemont, P. S. & Crumpton, M. J. (1991) Eur. J. Biochem. **198**, 749–760
75. Peretti, M., Becherruci, C., Mugridge, K. G., Solito, E., Silvestri, S. & Parente, L. (1991) Br. J. Pharmacol. **103**, 1327–1332
76. Hayashi, H., Owada, M. K., Sonobe, S., Domae, K., Yamanouchi, T., Kakunuga, T., Kitajima, Y. & Yaoita, H. (1990) Biochem. J. **269**, 709–715
77. Davidson, F. F., Lister, M. & Dennis, E. A. (1990) J. Biol. Chem. **265**, 5602–5609
78. Blackwood, R. A. & Ernst, J. D. (1990) Biochem. J. **266**, 195–200
79. Meers, P., Daleke, D., Hong, K. & Papahadjopoulos, D. (1991) Biochemistry **30**, 2903–2908
80. Bazzi, M. D. & Nelsuesten, G. L. (1991) Biochemistry, **30**, 971–979
81. Edwards, H. C. & Crumpton, M. J. (1991) Eur. J. Biochem. **198**, 121–129
82. Machoczek, K., Fischer, M. & Soling, H. D. (1989) FEBS Lett. **251**, 207–212

83. Toker, A., Ellis, C. A., Sellers, L. A. & Aitken, A. (1990) Eur. J. Biochem. **191**, 421–429
84. Lewit-Bentley, A. & Bodo, G. (1989) J. Mol. Biol. **210**, 875–876
85. Huber, R., Römisch, J. & Paques, E. (1990) EMBO J. **12**, 3867–3874
86. Huber, R., Schneider, M., Mayr, I., Römisch, J. & Paques, E. P. (1990) FEBS Lett. **275**, 15–21
87. Pietrobon, D., Di Virgilio, F. & Pozzan, T. (1990) Eur. J. Biochem. **193**, 599–622
88. Meyer, T. (1991) Cell **64**, 675–678
89. Hazen, S. L., Stuppy, R. J. & Gross, R. W. (1990) J. Biol. Chem. **265**, 10622–10630
90. Clark, J. D., Lin, L. L., Kriz, R. W., Ramesha, C. S., Sultzman, L. A., Lin, A. Y., Milona, N. & Knopf, J. L. (1991) Cell **65**, 1043–1045
91. Kramer, R. M., Roberts, E. F., Manetta, J. & Putnam, J. E. (1991) J. Biol. Chem. **266**, 5268–5272
92. Sharp, J. D., White, D. L., Chiou, G., Goodson, T., Gamboa, G. C., McClure, D., Burgett, S., Hoskins, J. A., Skatrud, P. L., Sportsman, J. R., Becker, G. W., Kang, L. H., Roberts, E. F. & Kramer, R. M. (1991) J. Biol. Chem. **266**, 14850–14853
93. Davidson, F. F. & Dennis, E. A. (1990) J. Mol. Evol. **31**, 228–238
94. Touqui, L., Rothhut, B., Shaw, A. M., Fradin, A., Vargaftig, B. B. & Russo-Marie, F. (1986) Nature (London) **321**, 177–180
95. Goldschmidt, P. J., Kim, J. W., Machesky, L. M., Rhee, S. G. & Pollard, T. D. (1991) Science **251**, 1231–1233
96. Cirino, G., Flower, R. J., Browning, J. L., Sinclair, L. K. & Pepinsky, R. B. (1987) Nature (London) **328**, 270–272
97. Errasfa, M., Bachelet, M. & Russo-Marie, F. (1988) Biochem. Biophys. Res. Commun. **153**, 1267–1270
98. Fradin, A., Rothhut, B., Poincelot-Canton, B., Errasfa, M. & Russo-Marie, F. (1988) Biochim. Biophys. Acta **963**, 248–257
99. Maridonneau-Parini, I., Errasfa, M. & Russo-Marie, F. (1989) J. Clin. Invest. **83**, 1936–1940
100. Solito, E. & Parente, L. (1989) Br. J. Pharmacol. **96**, 656–660
101. Cirino, G., Peers, S. H., Flower, R. J., Browning, J. L. & Pepinsky, R. B. (1989) Proc. Natl. Acad. Sci. U.S.A. **86**, 3428–3432
102. Errasfa, M. & Russo-Marie, F. (1989) Br. J. Pharmacol. **97**, 1051–1058
103. Wery, J. P., Schevitz, R. W., Clawson, D. K., Bobbitt, J. L., Dow, E. R., Gamboa, G., Goodson, T., Jr, Hermann, R. B., Kramer, R. M., McClure, D. B., Mihelich, E. D., Putnam, J. E., Sharp, J. D., Stark, D. H., Teater, C., Warrick, M. W. & Jones, N. D. (1991) Nature (London) **352**, 79–82
104. Scott, D. L., White, S. P., Browning, J. L., Rosa, J. J., Gelb, M. H. & Sigler, P. (1991) Science, **254**, 1007–1010
105. Russo-Marie, F., Paint, M. & Duval, D. (1979) J. Biol. Chem. **254**, 8498–8504
106. Solito, E., Raugie, G., Melli, M. & Parente, L. (1991) FEBS Lett. **291**, 238–244
107. Flower, R. (1988) Br. J. Pharmacol. **94**, 987–1015
108. Johnson, M. D., Kamso-Pratt, J., Pepinsky, R. B. & Whetsell, W. O. (1989) Am. J. Clin. Pathol. **92**, 424–430
109. Johnson, M. D., Kamso-Pratt, J., Pepinsky, R. B. & Whetsell, W. O. (1989) Hum. Pathol. **20**, 772–776
110. Smillie, F., Bolton, C., Peers, S. H. & Flower, R. J. (1989) Br. J. Pharmacol. **97**, 90–95
111. Strijbos, P. J. L. M., Tilders, F. J. H., Carey, F., Forder, R. & Rothwell, N. J. (1990) Brain Res. **553**, 249–260
112. Carey, F., Rorder, R., Edge, D., Grenne, A. R., Horan, P. J. L. M. & Rothwell, N. (1990) Am. J. Physiol. **259**, R266–R271
113. Davidson, J., Flower, R. J., Milton, A. S., Peers, S. H. & Rotondo, D. (1991) Br. J. Pharmacol. **102**, 7–9
114. Relton, J. K., Strijbos, P. J. L. M., O'Shaugnessy, C. T., Carey, F., Forder, R. A., Tilders, F. J. H. & Rothwell, N. J. (1991) J. Exp. Med. **174**, 305–310

# Evolutionary conservation and three-dimensional folding of the tyrosine kinase substrate annexin II

**Volker Gerke**

Department of Biochemistry, Max Planck Institute for Biophysical Chemistry, P.O. Box 2841, Am Fassberg, D-3400 Göttingen, Federal Republic of Germany

## Historical background

Annexin II was originally identified more than 10 years ago as a major cellular target of pp60$^{src}$, the tyrosine kinase encoded by the *src* oncogene of Rous sarcoma virus (RSV) [1–3]. Antibodies against gel-purified annexin II (at that time known as p36) revealed a submembraneous localization of this *src* kinase substrate which was similar but not identical to that of the enzyme itself (for reviews and references see [4–6]). Although it is a relatively abundant protein in chicken embryo fibroblasts where it was initially identified, it was not until 1984 that a purification of annexin II from chicken fibroblasts [7] or porcine intestinal epithelium [8] was reported. Both protocols led to the identification of a smaller polypeptide (now known as p11), which forms a tight complex with annexin II. The purification protocol from intestinal epithelium introduced as a highly selective step the enrichment of annexin II by $Ca^{2+}$-dependent binding to membrane structures and/or cytoskeletal elements followed by its solubilization through chelation of $Ca^{2+}$. This approach, which was independently developed for various other annexins (see other chapters of this volume), led to the preparation of milligram quantities of the annexin II–p11 complex (also called protein I or calpactin I) and thereby enabled a thorough biochemical characterization to be carried out [8,9].

Similar $Ca^{2+}$-dependent fractionation schemes yielded a 36 kDa protein as part of different annexin preparations which was also termed chromobindin 8 [10], lipocortin II [11] and placental anticoagulant protein IV [12]. Immunological and protein sequence analyses revealed that these proteins are indeed identical to the *src* kinase substrate isolated from chicken fibroblasts and porcine intestinal epithelial cells (for review on nomenclature see [13]).

Although the different names given to annexin II reflect different potential functions (e.g. in chromaffin granule fusion with the plasma membrane or inhibition of blood coagulation [12,14,15]), the precise role of this protein within the cell is still unknown. However, the fact that it represents a substrate not only for $pp60^{src}$ but also for different other signal-transducing protein kinases (see below) suggests that annexin II might be involved in processes of cellular growth and/or differentiation (for review see [16]). The detailed structural knowledge of annexin II, which has accumulated in the recent past and will be reviewed below, should enable us to develop more precise models of annexin II function and to test these models in the living cell.

## Biochemical characterization and phosphorylation

Some biochemical properties of annexin II, which were described for the protein purified from porcine [8] and later also bovine [9] intestinal epithelium, are summarized in Table 1. Once separated from its p11 subunit (achieved by gel filtration under denaturing conditions and subsequent renaturation), annexin II is a soluble monomeric molecule. As expected from the purification protocol, it binds in a $Ca^{2+}$-dependent manner to elements of the cytoskeleton (actin filaments and proteins of the spectrin family) and negatively charged phospholipids, in particular phosphatidylserine and phosphatidylinositol [9,17,18]. Spectroscopical analysis revealed a profound conformational change in annexin II, which occurs upon the addition of $Ca^{2+}$ and results in a decreased u.v.-absorption between 260 and 300 nm and a blue shift in fluorescence emission once the protein is excited at 295 nm [17,19,20]. Although relatively high $Ca^{2+}$ concentrations (0.5–1 mM) are required to saturate this effect and to establish the interaction with F-actin and non-erythroid-spectrins, a binding to phospholipid is already observed at less than 10 μM $Ca^{2+}$, arguing for a mutual influence of $Ca^{2+}$- and phospholipid-binding [9,21–23].

Binding of the p11 dimer to annexin II, which is solely mediated through the *N*-terminal 12 amino acid residues (see chapter 6 by Weber), leads to a dimerization of the annexin II chain [17,24,25]. The annexin $II_2p11_2$ tetramer, which is probably the predominant species in most cells, has a higher $Ca^{2+}$-affinity in the presence of phospholipid than the annexin II monomer. The complex is able to bind phospholipid vesicles at submicromolar $Ca^{2+}$ concentrations [23]. In addition, the heterotetramer

**Table 1. Biochemical properties and phosphorylation of annexin II**

| | Annexin II | Annexin $II_2p11_2$ |
|---|---|---|
| $M_2$ | 38 500* | 85 000† |
| Isoelectric point | 7.4 | |
| Stokes radius | 23 Å | 41 Å |
| Phospholipid-binding | | |
| PS | 0.65 μM [21]‡, 9 μM [9], > 10 μM [23] | < 0.01 μM [23] |
| PI | 1.3 μM [21] | |
| PA/PE | 0.2 μM [21] | |
| Phosphorylation | | |
| Tyrosine | *src* encoded tyrosine-kinase | *src* kinase |
| | *fps* encoded tyrosine-kinase | *fps* kinase |
| | Insulin receptor kinase | |
| Serine | Protein kinase C | Protein kinase C |
| | cAMP-dependent kinase | — |
| | CaM-dependent kinase | — |

* *Calculated from amino acid sequence;* † *native* $M_r$ *as determined by analytical ultracentrifugation;* ‡ *different* $Ca^{2+}$ *concentrations for half-maximal PS-binding of the annexin II monomer have been reported. See text for further references. PA, phosphatidic acid; PE, phosphatidylethanolamine; PI, phosphatidylinositol; PS, phosphatidylserine.*

but not monomeric annexin II is able to aggregate and even fuse chromaffin granules at $Ca^{2+}$ concentrations (0.7 μM) that parallel $Ca^{2+}$ requirements for secretion in permeabilized chromaffin cells [15]. Thus, p11 binding not only modulates the physical state of annexin II but also regulates at least some of its biochemical properties.

Limited proteolysis of monomeric and p11-complexed annexin II under native conditions yields a stable protein core that was shown by sequence analysis (see below and Fig. 1) to start at amino acid 14 (V8 protease digestion), 28 (trypsin digestion) and 30 (chymotrypsin digestion), respectively [26]. This approach and the biochemical characterization of the annexin II core led to the assignment of the binding sites for $Ca^{2+}$, phospholipid and F-actin to the core [9,18]. The *N*-terminal 29 amino acids (also known as the tail) contain not only the binding site for p11 (amino acids 1–12) [25,27] but also different phosphorylation sites. While

**Fig. 1. Sequence comparison of annexin II sequences from different species.**

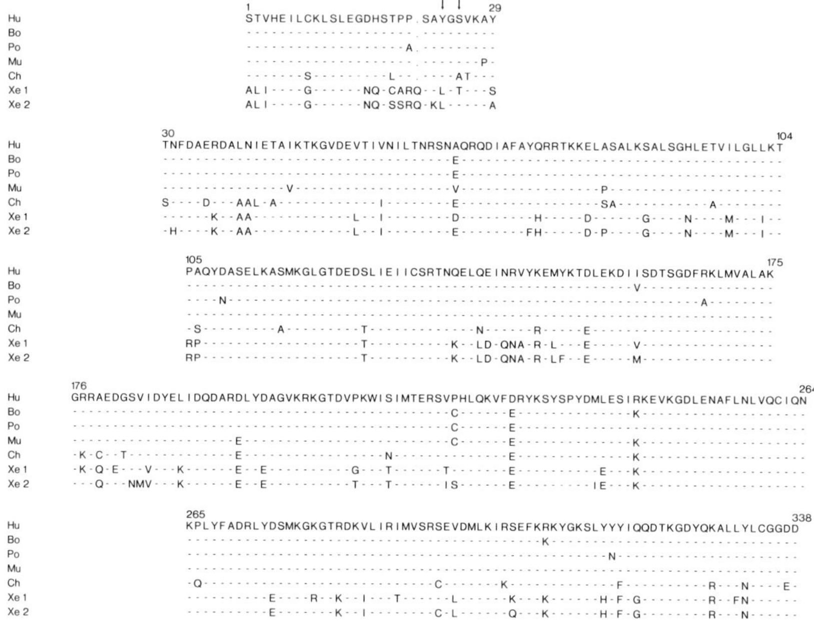

*Deduced amino acid sequences are given for human (Hu) [11], bovine (Bo) [51], porcine (Po; V. Gerke, unpublished work), mouse (Mu) [52], chicken (Ch) [54] and two types of* Xenopus laevis *annexin II (Xe) [55]. Residues identical to the sequence given at the top are indicated by dashes. The four annexin repeats (amino acids 30–104, 105–175, 176–264, 265–338) and the* N*-terminal tail (amino acids 1–29) are arranged separately. Sites for pp60$^{src}$ (Tyr-23) and protein kinase C phosphorylation (Ser-25) are marked by downward arrows.*

Tyr-23 is the target for pp60$^{src}$, Ser-25 is phosphorylated by protein kinase C [28–30]. This serine phosphorylation occurs upon phorbol 12-myristate 13-acetate-treatment of various cultured cells lines and is also observed *in vitro* [29,30]. Tyrosine-phosphorylation is observed *in vivo* in chicken embryo fibroblasts after RSV infection and in a RSV-transformed rat kidney cell line [1,2,8]. *In vitro*, annexin II is phosphorylated on tyrosine not only by pp60$^{src}$ [31] but also by the insulin receptor tyrosine kinase

[32] and the tyrosine kinase encoded by the *fps* oncogene of Fujinami sarcoma virus [33]. Although the tyrosine residue phosphorylated by the last two kinases has not been mapped precisely, Tyr-23 is the most likely candidate. Tyr-23 is also the site for a phosphorylation of annexin II by the epidermal growth factor (EGF) receptor-kinase, which is observed in human A431 cells upon EGF treatment [34]. On the basis of the various phosphorylation events it has been suggested that annexin II is involved in signal transduction pathways that regulate cellular growth and/or differentiation. However, direct experimental evidence for this hypothesis has yet to emerge as (i) the precise physiological role of annexin II is not known (see also other chapters of this volume), and (ii), tyrosine phosphorylation has only a marginal impact on the biochemical properties displayed by purified annexin II (phosphorylated annexin II has a somewhat lower affinity towards $Ca^{2+}$ and/or phospholipid than the unphosphorylated form [23]). In addition, only 10–15 % of the annexin II molecules are estimated to be phosphorylated in, for example, RSV-transformed chicken cells [3], and the intracellular distribution and properties of this fraction have not been analysed thoroughly.

Interestingly, the serine and tyrosine phosphorylation events appear mutually exclusive, i.e. Tyr-23-phosphorylated annexin II is not phosphorylated by protein kinase C on Ser-25 and *vice versa* [29]. This effect most likely results from structural perturbations that are introduced into this region of the molecule upon phosphorylation by either enzyme. They render this site inaccessible for the other enzyme but do not interfere with p11 binding. Both Tyr-23- and Ser-25-phosphorylated annexin II bind p11 and form the tetrameric annexin $II_2p11_2$ complex, indicating that these phosphorylations and p11-binding appear to modulate annexin II properties in different ways [23,30]. Two additional serine phosphorylation sites are present in the *N*-terminal tail of annexin II (amino acids 1–29), that are targets for calmodulin- and cAMP-dependent protein kinases [30]. Phosphorylation at these sites renders the annexin II molecule inactive in p11 binding. Thus, it seems possible that the physical state of annexin II within the cell, i.e. the ratio of monomeric to p11-bound annexin II, is regulated through phosphorylation events.

## Intracellular distribution and induction of annexin II expression upon cell differentiation

Within most differentiated cells annexin II is present in the cortical cytoplasm showing an intracellular distribution very similar to its *in vitro* ligand, non-erythroid-spectrin (for review and references see [4,6]). However, several experimental data argue against a direct interaction between annexin II and non-erythroid-spectrin *in vivo*. Microinjection of monoclonal antibodies against p11 leads to the formation of annexin II–p11 aggregates without affecting the non-erythroid-spectrin distribution [35], whereas the injection of spectrin antibodies leads to spectrin aggregates, without having a disturbing effect on annexin II [36]. Because of its interaction with annexin II, p11 assumes the same submembranous location [35,37]. The association of annexin II and p11 with the cytoplasmic face of the plasma membrane is dependent on the presence of $Ca^{2+}$ but still occurs at sub-micromolar $Ca^{2+}$ concentrations [38].

Both annexin II and p11 seem coordinately expressed and have been identified in a variety of cell lines and tissues. They are most abundant in intestinal epithelium, lung and placenta, but seem absent from or present at only very low levels in skeletal muscle, liver and erythrocytes [35,39–41]. However, it is interesting to note that at least annexin II is expressed in human hepatocellular carcinomas, whereas no expression is detected in normal hepatocytes [42].

Within the cell annexin II seems to exist in two different forms: the monomeric molecule, which seems if at all only loosely associated with the cortical cytoskeleton, and the complexed form (the annexin $II_2p11_2$ heterotetramer), which is anchored in the submembranous cytoskeleton [35]. The ratio of monomer:heterotetramer seems to vary in different cells; intestinal epithelial cells contain more than 90% in the complexed form [8], whereas chicken embryo fibroblasts may have a 50% excess of monomeric annexin II [7]. Although different intracellular functions have not been assigned to the different pools of annexin II, it seems possible that certain phosphorylations of annexin II modulate its intracellular activities by changing the monomer:heterotetramer ratio (see above).

A significant increase in annexin II expression has been detected during the differentiation of various cells, e.g. during the differentiation of mesenchymal cells into connective tissue and cartilage [40,43] and in cells of the differentiating chicken lens [44]. The synthesis of annexin II is also induced in different cell lines, which can be induced to differentiate *in vitro*. This has been shown for the annexin II mRNA in mouse fibroblasts after serum stimulation [45] and for the annexin II protein in human U937 cells (a myeloid cell line), which were induced to differentiate into macrophage-like cells by treatment with phorbol ester [46]. In addition, the level of annexin II protein in rat phaeochromocytoma PC12 cells is more than 10-fold higher after nerve growth factor (NGF) treatment, which induces differentiation into non-dividing neuron-like cells [39,47]. Interestingly, the synthesis of p11 mRNA is also induced upon NGF treatment of PC12 cells following a time course similar to that seen for annexin II induction [48]. The physiological consequences of these inductions are not known. It has been observed, however, that overexpression of p11 in undifferentiated PC12 cells (achieved by transfection with the p11 cDNA) can induce morphological changes reminiscent of the NGF-induced phenotype, i.e. the outgrowth of cell processes [49]. On the basis of comparisons of Northern and Western blot analyses it seems that the induction of annexin II as well as p11 expression occurs at the level of transcription. DNA elements which could mediate this induction by binding certain transcription factors have been identified in the annexin II promotor [50]. However, it has not yet been established that this is indeed the mode of regulation of annexin II transcription.

## Annexin II is highly conserved in different vertebrates

The first annexin II sequences were obtained by cloning the human [11], bovine [51] and murine [52] cDNA, and by partial protein sequencing of the porcine molecule [53]. Analysis of the primary structure revealed a high degree of conservation and the typical domain structure characterized by the four so-called annexin repeats. Subsequently, cDNAs for porcine (V. Gerke, unpublished work), chicken [54] and *Xenopus laevis* [55] annexin II have been isolated. A comparison of all deduced amino acid sequences is depicted in Fig. 1. It emphasizes the repeat structure of the annexin II core (amino acids 30–338) with the individual repeats (the units showing inter- as well as intramolecular sequence similarity in the annexin family) covering amino acids 30–104 (repeat 1), 105–175 (repeat 2), 176–264 (repeat 3) and 265–338 (repeat 4). The *N*-terminal tail (residues 1–29), which is unique among the annexin family, is written separately. The high evolutionary conservation is particularly evident in the core domain which harbours the binding sites for the common annexin ligands ($Ca^{2+}$, phospholipids). In addition, the very *N*-terminal 14 amino acids, i.e. the binding site for p11, are almost invariant among the different species. The sole amino acid exchanges in this region are found at positions 1, 2 and 3 of the *X. laevis* molecules and at position 8. The replacements occurring do not appear to affect the ability of the protein to interact with p11, as *X. laevis* annexin II binds human p11 in a gel overlay experiment [55]. Thus, the p11 binding site is probably a separate functional domain which is highly conserved throughout evolution. Evidence for this hypothesis also stems from the fact that this domain is encoded by a separate exon in the annexin II gene [50,56].

The highest degree of sequence divergence between the different annexin II species is observed between amino acids 15 and 29, i.e. in the connector region between the p11 binding domain and the protein core. These sequence variations as well as the high susceptibility towards proteolysis [9,26] indicate that this part of the molecule has a less defined secondary and tertiary structure. The connector region is, however, of regulatory importance because it contains the phosphorylation sites for pp60$^{src}$ (Tyr-23) and protein kinase C (Ser-25), and phosphorylation is known to modulate biochemical properties displayed by annexin II *in vitro* (see above). The clawed toad *X. laevis* expresses two annexin II species that are highly similar but are encoded by different genes (probably a result of chromosome duplication by tetraploidization [55]). Interestingly, both annexin II molecules from *X. laevis* contain a leucine-for-tyrosine replacement at position 23 which eliminates the pp60$^{src}$ phosphorylation site. It remains to be seen whether properties of *X. laevis* annexin II are regulated by protein kinase C phosphorylation, as an appropriate site (Ser-25) is present in one *X. laevis* molecule.

## Three-dimensional folding

The first member of the annexin family whose crystal structure has been determined is annexin V (see chapter 10 by Huber *et al.*) [57,58]. A highly symmetric molecule with densely packed α-helices was revealed and the four annexin repeats could be detected as separate domains in the crystal. Repeats 1 and 4 and repeats 2 and 3 form tightly associated modules that are stabilized by hydrophobic inter-repeat contacts. The two modules (repeats 1/4 and 2/3 respectively) interact more loosely with one another. This interaction is mediated through polar side chains and yields a central pore that transverses the molecule and most likely is responsible for the observed $Ca^{2+}$-channel activity of annexin V [59]. Given the high sequence homology and the fact that all other annexins are characterized by the typical repeats, it is likely that the other members of the family show a similar three-dimensional folding. Therefore, I will discuss aspects of the annexin II structure in light of the annexin V structure and recent biochemical data that describe molecular characteristics of a discontinuous epitope and a $Ca^{2+}$-binding site in annexin II.

If the annexin V structure serves as a blueprint, the position of the five α-helices in each repeat of annexin II can be drawn as depicted in Fig. 2. Typically the endonexin fold, a sequence of 17 amino acids which is present in each repeat and highly conserved among the annexins [60], forms the loop between helices *a* and *b* as well as helix *b* [57]. In the annexin V crystals, $Ca^{2+}$-binding sites, which are present in repeats 1, 2 and 4, are formed by the loops of these endonexin folds [58]. Additional coordination of the $Ca^{2+}$-ion is provided by the carboxyl oxygens of a glutamic acid or aspartic acid located in helix *d* in each repeat; a schematic drawing of this annexin-type $Ca^{2+}$-binding site is shown in Fig. 3. A similar architecture can be expected for $Ca^{2+}$ sites in annexin II since the loop as well as the acidic residue in helix *d* are almost invariant in repeats 2 and 4. However, the situation is different in repeats 1 and 3. Two amino acid substitutions in the loop between helices *a* and *b* (Thr-47 for a glycine and Val-50 for a threonine) as well as an alanine replacement for the glutamic acid in helix *d* might render this $Ca^{2+}$ site in repeat 1 inactive. In repeat 3, sequence differences in the *C*-terminal portion of helix *a* as well as in the loop between helices *a* and *b* might cause a different folding in this region which is also evident from secondary structure predictions [61]. In addition, an acidic residue (Glu-246) is present in the loop between helices *d* and *e*, which in the primary structure has the appropriate distance to the loop between helices *a* and *b* to serve as a $Ca^{2+}$-coordination site. The presence of an intact $Ca^{2+}$-binding site in repeat 3 of annexin II has also been demonstrated experimentally [62]. These studies made use of a unique tryptophan in annexin II (Trp-212 located in repeat 3) whose fluorescence emission is significantly altered by $Ca^{2+}$-binding in close proximity [19]. If conserved residues in the interhelical loop in repeat 3 (Gly-206 and Thr-207) are changed into alanine by site-directed mutagenesis, the $Ca^{2+}$ response in Trp-212 fluorescence emission is significantly impaired [62]. Interestingly, these mutations also reduce the affinity of the entire protein

**Fig. 2. Proposed structural organization of annexin II.**

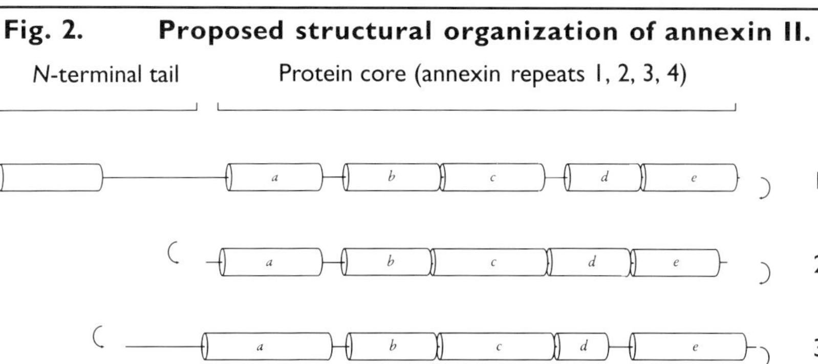


*The schematic illustration is based on the crystal structure determined for annexin V [57]. The predicted α-helices are depicted as cylinders (helices* a–e *in each annexin repeat and the* N-*terminal helix that harbours the p11 binding site).*

towards $Ca^{2+}$ and/or phospholipid indicating that some cooperation might exist between the different $Ca^{2+}$- and/or phospholipid-binding sites. Thus, three $Ca^{2+}$ sites of the annexin type seem to be present in both annexin V and annexin II. However, whereas these sites are located in

**Fig. 3. Schematic drawing of an annexin-type $Ca^{2+}$-binding site.**

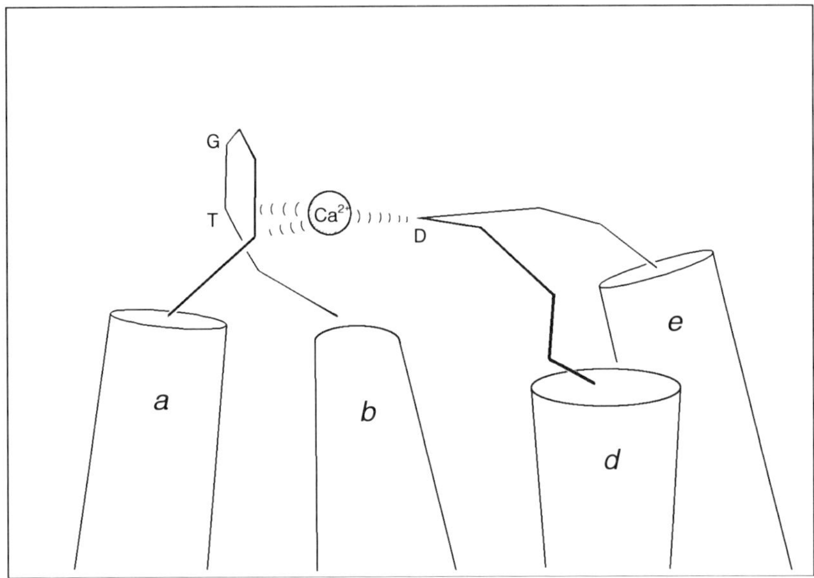


*Carbonyl oxygens of the loop between helix* a *and* b*, which includes the conserved glycine (G) and threonine (T) residues, and carboxyl oxygens of a conserved acidic residue (D) located in helix* d *or in the loop between helices* d *and* e *provide coordination of the $Ca^{2+}$ ion. Alpha-helices are depicted as cylinders. The model is based on the structure of $Ca^{2+}$-binding sites in human annexin V [58].*

repeats 1, 2 and 4 in annexin V, repeats 2, 3 and 4 seem to display $Ca^{2+}$ binding in annexin II.

Because of the divergent *N*-terminal tails, the three-dimensional folding in this region of the annexin II molecule cannot be extrapolated from the annexin V structure. Important information as to the position of the regulatory connector region relative to the protein core stems from the mapping of a three-dimensional epitope on annexin II. Using limited proteolysis [26] and *in vitro* mutagenesis [63] it was shown that this epitope involves Lys-27 in the connector region as well as Arg-62, Glu-65 and

Arg-67 which reside in helices *b* and *c* of repeat 1. In the annexin V crystal the corresponding residues are located on the concave side of the molecule whereas the bound $Ca^{2+}$ ions are found exclusively on the opposite convex side [57,58]. This spatial distance raises questions about the mechanism of regulation of annexin II properties by phosphorylation. The phosphorylation sites (Tyr-23 and Ser-25) are located in the connector region in close proximity to the *N*-terminal part of the epitope. Nevertheless, Tyr-23 phosphorylation by $pp60^{src}$ (at the concave side of annexin II) leads to an enhanced $Ca^{2+}$ requirement in phospholipid-binding, and thus modulates events occurring at the convex side of the protein [23]. One possible explanation for this phenomenon would be a phosphorylation-induced conformational change in annexin II which leads to altered $Ca^{2+}$/lipid-binding properties. Possibly, this signal is transmitted through the molecule via the close contact of the phosphorylated residues in the connector region and the above-mentioned charged residues in helices *b* and *c* of repeat 1.

The structural basis of the regulation of annexin II properties by p11-binding is not yet clear. The p11-induced formation of the annexin $II_2p11_2$ heterotetramer produces a markedly increased affinity for $Ca^{2+}$ and/or phospholipid [15,23]. Interestingly, the annexin II core also has a higher affinity for these ligands than the intact annexin II monomer. Thus, the *N*-terminal tail which includes the p11-binding site seems to have an inhibitory effect on $Ca^{2+}$/lipid-binding. Because the position of the *N*-terminal portion of the tail relative to the protein core has not been determined so far, it remains possible that the inhibitory effect is either transmitted through conformational changes or is the consequence of some spatial interference (such as the p11-binding site folding back on the convex side of the molecule, i.e. the side harbouring the $Ca^{2+}$-binding sites). In addition, p11-binding could also influence annexin II properties by simply doubling the binding sites for $Ca^{2+}$ and phospholipid (in the heterotetramer as compared to the annexin II monomer). Analysis of the crystal structure of the entire annexin II molecule and the annexin $II_2p11_2$ complex should settle these questions.

*I thank Klaus Weber for critical reading of the manuscript and Carsten Thiel for help in preparing the figures.*

## References

1. Erikson, E. & Erikson, R. L. (1980) Cell **21**, 829–836
2. Radke, K. & Martin, G. S. (1979) Proc. Natl. Acad. Sci. U.S.A. **76**, 5212–5216
3. Radke, K., Gilmore, T. & Martin, G. S. (1980) Cell **21**, 821–828
4. Campos-Gonzáles, R. & Glenney, J. R. Jr. (1990) in Stimulus Response Coupling: The Role of Intracellular Calcium-binding Proteins (Smith, V. L. & Dedman, J. R., eds.), pp. 339–356, CRC Press, Boca Raton
5. Cooper, J. A. & Hunter, T. (1983) Curr. Top. Microbiol. Immunol. **107**, 125–161

6. Gerke, V., Johnsson, N. & Weber, K. (1990) in Stimulus Response Coupling: The Role of Intracellular Calcium-binding Proteins (Smith, V. L. & Dedman, J. R., eds.), pp. 311–338, CRC Press, Boca Raton
7. Erikson, E., Tomasiewicz, H. G. & Erikson, R. L. (1984) Mol. Cell. Biol. **4**, 77–85
8. Gerke, V. & Weber, K. (1984) EMBO J. **3**, 227–233
9. Glenney, J. R. Jr. (1986) J. Biol. Chem. **261**, 7247–7252
10. Creutz, C. E., Zaks, W. J., Hamman, H. C., Crane, S., Martin, W. H., Gould, K. L., Oddie, K. M. & Parsons, S. J. (1987) J. Biol. Chem. **262**, 1860–1868
11. Huang, K.-S., Wallner, B. P., Mattaliano, R. J., Tizard, R., Burne, C., Frey, A., Hession, C., McGray, P., Sinclair, L. K., Chow, E. P., Browning, J. L., Ramachandran, K. L., Tang, J., Smart, J. E. & Pepinsky, R. B. (1986) Cell **46**, 191–199
12. Tait, J. F., Sakata, M, McMullen, B. A., Miao, C. H., Funakoshi, T., Hendrickson, L. E. & Fujikawa, K. (1988) Biochemistry **27**, 6268–6276
13. Moss, S. E., Edwards, H. C. & Crumpton, M. J. (1991) in Novel Calcium-binding Proteins (Heizmann, C. W., ed.), pp. 535–566, Springer Verlag, Berlin
14. Ali, S. M., Geisow, M. J. & Burgoyne, R. D. (1989) Nature (London) **340**, 313–315
15. Drust, D. S. & Creutz, C. E. (1988) Nature (London) **331**, 88–91
16. Gerke, V. (1989) Cell Motil. Cytoskeleton **14**, 449–454
17. Gerke, V. & Weber, K. (1985) J. Biol. Chem. **260**, 1688–1695
18. Johnsson, N., Vandekerckhove, J., van Damme, J. & Weber, K. (1986) FEBS Lett. **198**, 361–364
19. Marriott, G., Kirk, W. R., Johnsson, N. & Weber, K. (1990) Biochemistry **29**, 7004–7011
20. Pigault, C., Follenius-Wund, A., Lux, B. & Gerard, D. (1990) Biochim. Biophys. Acta **1037**, 106–114
21. Blackwood, R. A. & Ernst, J. D. (1990) Biochem. J. **266**, 195–200
22. Davidson, F. F., Dennis, E. A., Powell, M. & Glenney, J. R. (1987) J. Biol. Chem. **262**, 1698–1705
23. Powell, M. A. & Glenney, J. R. (1987) Biochem. J. **247**, 321–328
24. Glenney, J. R. Jr., Boudreau, M., Gaylean, R., Hunter, T. & Tack, B. (1986) J. Biol. Chem. **261**, 10485–10488
25. Johnsson, N., Marriott, G. & Weber, K. (1988) EMBO J. **7**, 2435–2442
26. Johnsson, N., Johnsson, K. & Weber, K. (1988) FEBS Lett. **236**, 201–204
27. Becker, T., Weber, K. & Johnsson, N. (1990) EMBO J. **9**, 4207–4213
28. Glenney, J. R. Jr. & Tack, B. F. (1985) Proc. Natl. Acad. Sci. U.S.A. **82**, 7884–7888
29. Gould, K. L., Woodgett, J. R., Isacke, C. M. & Hunter, T. (1986) Mol. Cell. Biol. **6**, 2738–2744
30. Johnsson, N., Van, P. N., Söling, H.-D. & Weber, K. (1986) EMBO J. **5**, 3455–3460
31. Glenney, J. R. Jr. (1985) FEBS Lett. **192**, 79–82
32. Karasik, A., Pepinsky, R. B., Shoelson, S. E. & Kahn, C. R. (1988) J. Biol. Chem. **263**, 11862–11867
33. Hagiwara, M., Ochiai, M., Owada, K., Tanaka, T. & Hidaka, H. (1988) J. Biol. Chem. **263**, 6438–6441
34. Isacke, C. M., Trowbridge, I. S. & Hunter, T. (1986) Mol. Cell. Biol. **6**, 2745–2751
35. Zokas, L. & Glenney, J. R. Jr. (1987) J. Cell Biol. **105**, 2111–2121
36. Mangeat, P. H. & Burridge, K. (1984) J. Cell Biol. **98**, 1363–1377
37. Osborn, M., Johnsson, N., Wehland, J. & Weber, K. (1988) Exp. Cell Res. **175**, 81–96
38. Semich, R., Gerke, V., Robenek, H. & Weber, K. (1989) Eur. J. Cell Biol. **50**, 313–323
39. Gould, K. L., Cooper, J. A. & Hunter, T. (1984) J. Cell Biol. **98**, 487–497
40. Greenberg, M. E., Brackenbury, R. & Edelman, G. M. (1984) J. Cell Biol. **98**, 473–486
41. Saris, C. J. M., Kristensen, T., D'Eustachio, P., Hicks, L. J., Noonan, D. J., Hunter, T. & Tack, B. F. (1987) J. Biol. Chem. **262**, 10663–10671
42. Frohlich, M., Motte, P., Galvin, K., Takahashi, H., Wands, J. & Ozturk, M. (1990) Mol. Cell Biol. **10**, 3216–3223
43. Carter, V. C., Howlett, A. R., Martin, G. S. & Bissel, M. J. (1986) J. Cell Biol. **103**, 2017–2024

44. Talian, J. C. & Zelenka, P. S. (1991) Dev. Biol. **143**, 68–77
45. Keutzer, J. C. & Hirschhorn, R. R. (1990) Exp. Cell Res. **188**, 153–159
46. Isacke, C. M., Lindberg, R. A. & Hunter, T. (1989) Mol. Cell. Biol. **9**, 232–240
47. Schlaepfer, D. D. & Haigler, H. T. (1990) J. Cell Biol. **111**, 229–238
48. Masiakowski, P. & Shooter, E. M. (1988) Proc. Natl. Acad. Sci. U.S.A. **85**, 1277–1281
49. Masiakowski, P. & Shooter, E. M. (1990) J. Neurosci. Res. **27**, 264–269
50. Spano, F., Rangei, G., Palla, E., Colella, C. & Melli, M. (1990) Gene **95**, 243–251
51. Kristensen, T., Saris, C. J. M., Hunter, T., Hicks, L. J., Noonan, D. J., Glenney, J. R. Jr. & Tack, B. F. (1986) Biochemistry **25**, 4497–4503
52. Saris, C. J. M., Tack, B. F., Kristensen, T., Glenney, J. R. Jr. & Hunter, T. (1986) Cell **46**, 201–212
53. Weber, K. & Johnsson, N. (1986) FEBS Lett. **203**, 95–98
54. Gerke, V. & Koch, W. (1990) Nucleic Acids Res. **18**, 4246
55. Gerke, V., Koch, W. & Thiel, C. (1991) Gene **104**, 259–264
56. Amiguet, P., D'Eustachio, P., Kristensen, T., Wetsel, R. A., Saris, C. J. M., Hunter, T., Chaplin, D. D. & Tack, B. F. (1990) Biochemistry **29**, 1226–1232
57. Huber, R., Römisch, J. & Paques, E.-P. (1990) EMBO J. **9**, 3867–3874
58. Huber, R., Schneider, M., Mayr, I., Römisch, J. & Paques, E.-P. (1990) FEBS Lett. **275**, 15–21
59. Rojas, E., Pollard, H. B., Haigler, H. T., Parra, C. & Burns, A. L. (1990) J. Biol. Chem. **265**, 21207–21215
60. Geisow, M. J., Fritsche, U., Hexham, J. M., Dash, B. & Johnson, T. (1986) Nature (London) **320**, 636–638
61. Barton, G. J., Newman, R. H., Freemont, P. S. & Crumpton, M. J. (1991) Eur. J. Biochem. **198**, 749–760
62. Thiel, C., Weber, K. & Gerke, V. (1991) J. Biol. Chem. **266**, 14732–14739
63. Thiel, C., Weber, K. & Gerke, V. (1991) FEBS Lett. **285**, 59–62

# Annexin II: interaction with p11

**Klaus Weber**

Max Planck Institute for Biophysical Chemistry,
Department of Biochemistry, P.O. Box 2841, D-3400 Göttingen,
Federal Republic of Germany

## General properties

When annexin II is purified from various sources under exclusion of protease action it is obtained as a complex with the small polypeptide p11 [1–6]. Monomeric annexin II when obtained in a different fraction could either be the intact molecule or a derivative truncated at the *N*-terminus and therefore unable to bind p11 (see below).

Protein sequence analysis has identified p11 as a member of the S-100 protein family [4,7,8], which has in the meantime acquired several additional members (Fig. 1). All of these molecules are built from two consecutive EF-hands, followed by unique *C*-terminal extensions some 10 to 30 residues in length [9]. From sequence data and direct experiments carried out with several S-100 proteins, these molecules are known to be $Ca^{2+}$-binding proteins [10,11]. The p11 protein differs from the rest of the S-100 family, in that the sequences of the loops are modified; the first loop is shortened by deletion of three residues and the second loop is modified by some unusual amino acid exchanges. The resulting EF-hands lack the features necessary for $Ca^{2+}$ binding ([4]; see also Fig. 1). Lack of $Ca^{2+}$ binding by p11 has also been verified experimentally [4,8]. Consequently, p11 interacts constitutively with its cellular target annexin II, whereas the other members of the family are thought to find their protein ligands only in the presence of $Ca^{2+}$. Like other S-100 proteins, p11 is a dimer as judged by crosslinking experiments [3,4] and fluorescence spectroscopy [12]. The dissociation constant for the dimer is around 100 nM [12].

For two other members of the S-100 protein family, target proteins are indicated under rather special *in vitro* conditions; S-100 binds to aldolase [13] while calcyclin binds glyceraldehyde-3-phosphate dehydrogenase [14]. In comparison the tetrameric p11–annexin II complex is more fully characterized. The stoichiometry of two copies each of p11 and annexin II deduced from gel patterns [3] was confirmed by fluorescence spectroscopy [12]. The latter procedure also estimated the dissociation constant to be well below 30 nM [12]. In the heterotetramer each p11 chain

**Fig. 1. Alignment of the amino acid sequence of p11 with other members of the S-100 protein family.**

```
                1                                                  49
p11             PSQMEHAMETMMFTFHKFA---GDKGYLTKEDLRVLMEKEFPGFLENQKDPL
S-100α          GSELETAMETLINVFHAHSGKEGDKYKLSKKELKELLQTELSGFLDAQKDAD
S-100β           SELEKAVVALIDVFHQYSGREGDKHKLKKSELKELINNELSHFLEEIKEQE
18A2            ARPLEEALDVIVSTFHKYSGKEGDKFKLNKTELKELLTRELPSFLFKRTDEA
MRP 14      TCKMSQLERNIETIINTFHQYSVKLGHPDTLNQGEFKELVRKDLQNFLKKENKNEK
2A9             ACPLDQAIGLLVAIFHKYSGREGDKHTLSKKELKELIQKEL-TIGSKLQD-A
MRP 8           LTELEKALNSIIDVYHKYSLIKGNFHAVYRDDLKKLLETECPQYI-RKK---

                50                                             96
p11             AVDKIMKDLDQCRDGKVGFQSFFSLIAGLTIACNDYF-VVHMKQKGKK
S-100α          AVDKVMKELDEDGDGEVDFQEYVVLVAALTVACNNFFWENS
S-100β          VVDKVMETLDSDGDGECDFQEFMAFVAMITTACHEFFEHE
18A2            AFQKVMSNLDSNRDNEVDFQEYCVFLSCIAMMCNEFFEGCPDKEPRKK
MRP 14          VIEHIMEDLDTNADKQLSFEEFIMLMARLTWASHEKMHEGDEGPGHHHKPGLGEGTP
2A9             EIARLMEDLDRNKDQEVNFQEYVTFLGALALIYNEALKG
MRP 8           GADVWFKELDINTDGAVNFQEFLILVIKMGVAAHKKSHEESHKE
```

*Sequences are from porcine p11 [4], bovine S-100α [34], bovine S-100β [35], murine 18A2 [37], human calcyclin or 2A9 [38] and human MRP8 and MRP14 [39]. Dashes indicate gaps necessary for optimal alignment. The predicted $Ca^{2+}$-binding loops of the two consecutive EF-hands are indicated by lines above the sequences. Note in the p11 sequence a gap of three residues in the first loop and an unusual serine residue at the C-terminal side of the second loop. In contrast to the other members of the S-100 family of proteins, p11 does not bind calcium (see text). MRP8 is identical to the CF antigen, a protein present at elevated levels in the serum of patients suffering from cystic fibrosis [40]. Residues identical in p11 and the different S-100-like proteins are given by bold letters.*

of the dimer binds to one annexin chain. In addition to these *in vitro* results, there is strong evidence for an *in vivo* complex formation: from immunofluorescence microscopy studies of various cultured cells and of tissue sections, the distribution of p11 is seen to coincide with that of annexin II [15,16]. The submembraneous location of the complex and its $Ca^{2+}$-dependent association with the cytoplasmic face of the plasma membrane was also confirmed by immuno-gold electron microscopy [17]. Although the true physiological role of the annexins is still unclear, p11 is a specific modulator of the properties displayed by annexin II. The conversion of the monomeric annexin II into the heterotetramer by p11 leads to a lower $Ca^{2+}$ requirement in phospholipid binding [18]; this is also reflected in the assay based on the aggregation of chromaffin granules [19].

## The *N*-terminal α-helix of annexin II binds p11

Mild chymotryptic treatment of the annexin II–p11 heterotetramer (or free annexin II) defined two domains of annexin II. The digest produces p11, the core domain of annexin II and a few peptides comprising the *N*-terminal region, the so-called tail of annexin II. The core starts with Thr-30 and retains $Ca^{2+}$-, lipid- and F-actin-binding properties but lacks the ability to form a complex with p11. Because a synthetic peptide corresponding to residues 1–29 of the annexin, including the N-acetyl group on Ser-1, inhibits the reassociation of the intact annexin and p11, interest has focused on the tail domain [20,21]. In view of the fact that a longer core of annexin II obtained by V8 protease starts at Gly-14 and also lacks p11 binding [21], residues 1–13 seem to be particularly important. A more direct approach to characterizing the p11-binding site has become possible using fluorescence spectroscopy.

Porcine annexin II treated with a 1.5-fold M excess of Acrylodan incorporates 0.7 mol of the fluorophore Prodan, which is present exclusively on Cys-8. Prodan-labelled annexin II retains the ability to associate with p11. Upon complex formation, the maximum-emission wavelength is shifted from 518 to 489 nm with a substantial narrowing in the fluorescence band width [12]. These changes have been used to delineate the p11 site on the tail domain. Fluorescently labelled peptides from a chymotryptic or tryptic treatment of annexin II were found to consist of residues 1–10 and 1–9, respectively. Fluorescence spectroscopy showed that both peptides bind to p11. For the latter peptide, binding was also demonstrated by gel-filtration analysis. The emission spectra of the complexes formed between p11 and the Prodan-labelled annexin II or its *N*-terminal peptides are nearly superimposable.

The first 18 residues of the tail were synthesized either without the *N*-terminal N-acetyl group (N-1–18) or with the natural blocking group (Ac-1–18). Binding experiments showed that one p11 dimer can bind up to two *N*-terminal peptides. The dissociation constant for the complex of a single N-1–18 peptide to the p11 dimer is about 20 μM, while Ac-1–18 and annexin II have 1000-fold higher affinities with $K_d$ values below 30 nM. Thus, the N-acetyl group on Ser-1 is a functional part of the p11-binding site.

Secondary-structure prediction rules indicate α-helical potential for the *N*-terminal 18 residues of annexin II, and circular dichroism spectroscopy shows that Ac-1–18 has a random conformation in normal buffer but assumes a helical configuration upon addition of trifluoroethanol. This conversion is completed at 50 % trifluoroethanol and yields 63 % α-helix, i.e. about 12 residues. A similar level of helix induction

**Fig. 2. Helical wheel presentation of the *N*-terminal 14 residues of annexin II.**

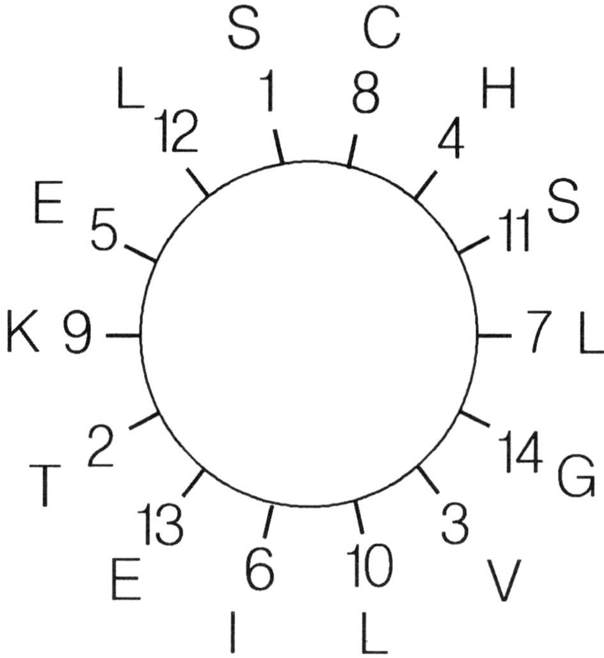


*This sequence comprises the p 11-binding site [12,22]. Note the hydrophobic face of the amphiphatic helix covering residues 3, 6, 7 and 10. For details see text.*

occurs when Ac-1–18 is complexed with p11. With the helical character of the binding site established experimentally, the structural motif is suggested by projecting the sequence of Ac-1–18 on a helical wheel [12]. The sequence indicates the features of an amphiphatic helix. The hydrophobic face comprises Val-3, Ile-6, Leu-7 and Leu-10. Charged residues locate to the opposite face, which also includes Leu-12 and Cys-8 (Fig. 2).

A competitive assay based on the complex between p11 and the fluorescently labelled *N*-terminal 14 residues of annexin II carrying the N-acetyl group has been used to refine the analysis [22]. The results showed that the binding site for p11 is fully contained in these 14 residues and most likely involves only the first 12 residues. Use of a large number of synthetic tetradecapeptides, which differ from the original binding sequence by single amino acid substitutions, confirms the hypothesis that an amphiphatic α-helix is induced upon binding to p11. The apparent affinities of the mutant peptides revealed that the N-acetyl group of Ser-1 and the hydrophobic side chains at positions 3, 6, 7 and 10 contribute most to the binding. The observed destabilization of the complex upon removal of single methyl groups from the hydrophobic face of the helix compares well with the known destabilization of proteins in which methyl groups are removed from the inner core. This led to the proposal that upon binding to p11 the hydrophobic side of the amphiphatic helix becomes fully buried. Although the strong contribution of the N-acetyl group is not clearly understood, current results [22] argue for a closely packed interaction between this group and the interace on p11.

The best studied member of the superfamily of proteins on the basis of consecutive EF-hands is calmodulin. Interestingly, calmodulin also binds to its target proteins via a short amphipathic helix. The use of myosin light chain kinase reveals this helix to be located at the *C*-terminal end [23–25]. A recently proposed model for the interaction of calmodulin and troponin C with amphipathic helices agrees with the general view. The hydrophobic sides of these two helices fit into a hydrophobic pocket, which seems to be a characteristic feature of most EF-hand molecules [26]. (For a more detailed discussion, see elsewhere [12,22].) The amphiphatic α-helix of annexin II, which from biochemical studies is known to cover as a functional unit residues 1–12 or 14, also appears as a separate unit in the gene [27]. The corresponding exon extends to residue 15.

## Attempts to characterize the annexin II binding site on p11

Although the binding site of p11 to the *N*-terminal α-helix (residues 1–12) of annexin II seems well established, far less is known about the contacts on p11. Currently, the most direct approach to elucidating this relies on a differential reactivity of the two cysteine residues of p11 under mild conditions with reagents such as 4-vinylpyridine, 2,2′-dithiopyridine and iodoacetamide [28]. In free p11 both Cys-61, located in the loop of the second EF-hand, and Cys-82, present in the *C*-terminal extension, are readily modified. The resulting p11 derivative shows an unchanged α-helical circular dichroism spectrum and retains the dimeric state but can no longer form a complex with annexin II. In contrast, in the heterotetramer, only Cys-61 is modified and the complex remains stable. The p11 prepared from this complex can again bind annexin II unless in a further reaction

Cys-82 is also modified. Thus, Cys-82 is located within or very near an important p11–annexin II contact region [28]. This study also provided a new method of preparing p11 from the heterotetramer without the need for denaturation by 9 M urea or 6 M guanidinium-HCl [3,8].

**Fig. 3. Comparison of p11 sequences from different vertebrates.**

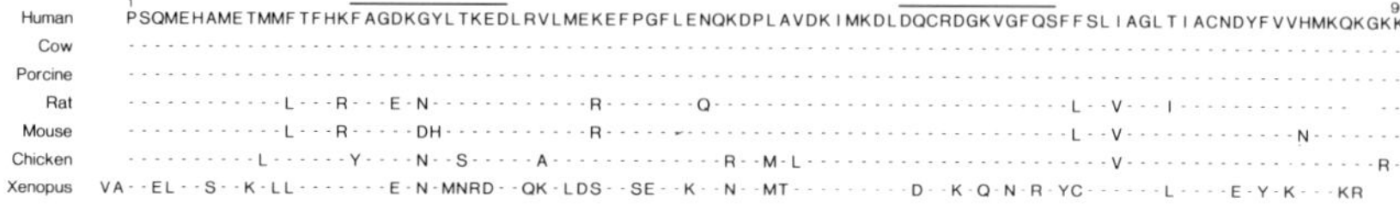

*Sequences are from man [29], cow [41], pig [4], mouse [41], rat [42], chicken [29] and* Xenopus laevis *[29]. Residues identical with the sequence given at the top are marked by dashes. Lines above the sequences mark the positions of the two EF-hand loops inactive in calcium binding. Note the few regions in which the* Xenopus *p11 retains sequence identity. One of them is the C-terminal extension following the second EF-hand.*

Additional clues to the contact points in p11 can be obtained by examining its interaction with a series of mutant annexin II peptides (see above). As described earlier, hydrophobicity seems to be a major stabilizing factor of the annexin II–p11 complex, and the interface between the annexin helix and p11 seems closely packed [22]. Comparison of several p11 sequences (Fig. 3) shows that various mammalian p11 molecules are very similar to chicken p11 (at least 90% identity), whereas p11 from *Xenopus laevis* exhibits a considerable degree of divergence (60% identity). Interestingly, the *C*-terminal 18 residues, which include Cys-82, show a relatively high conservation across the different species. Given the unique sequence of the *C*-terminal extension of p11 among the members of the S-100 protein family, some of its hydrophobic residues may be in contact with the annexin II helix [29].

These results have obvious implications for p11-related molecules with a *C*-terminal extension past the second EF-hand [28]. Four of them (S-100α, S-100β, S-100L and 18A2) share the cysteine corresponding to position 82 of p11 (see Fig. 1). If they exert their function as $Ca^{2+}$-controlled regulators, one can speculate that their *C*-terminal extensions might be involved in the binding to their yet unidentified targets. The lack of such an extension in the intestinal vitamin D-dependent calcium-binding protein would only emphasize its proposed function as a $Ca^{2+}$ buffer [30]. In contrast to these molecules, p11 has defective EF-hands and is not controlled by $Ca^{2+}$. Thus, it has to be in a permanent 'on' state in regard to its binding to annexin II (see above).

## Outlook

Although the physiological function of the annexins is not yet established, the recent X-ray analysis of annexin V [31,32] has created much excitement (see chapter 10 by Huber *et al.*). Analysis of the core domain reveals a channel or pore through the centre of the molecule, which may be a voltage-dependent $Ca^{2+}$ channel. The $Ca^{2+}$/phospholipid-binding sites locate to the convex upper side while the *N*-terminal tail is at the concave underside of the molecule. The tail domain of annexin II exceeds that of annexin V by 18 residues, with the protein kinase C sites being aligned (Thr-7 in annexin V, Ser-25 in annexin II; chapter 5 by Gerke). Thus, the exact positioning of the *N*-terminal amphipathic helix of annexin II versus the core domain remains to be elucidated. Nevertheless, the *N*-terminal p11-binding site as well as the phosphorylation sites (Tyr-23, Ser-25) could modulate $Ca^{2+}$/phospholipid-binding at the opposite side of the domain by varying the extent to which they are associated with the pore. The position of a discontinuous epitope of annexin II also favours this view [33].

One of the attractive feature of the annexin II–p11 complex has been the combination of members from two distinct protein families. This still leaves the puzzle of whether annexin II is just an oddity in the annexin family, or whether some of the other annexins can also bind a specific protein ligand, possibly a member of the S-100 family or an entirely distinct protein. At least the annexins with longer *N*-terminal tail domains are obvious candidates for having this property [12].

*I thank my colleagues T. Becker, V. Gerke and E. Kube for the figure material, and appreciate helpful discussions with V. Gerke and M. Osborn.*

### References

1. Erikson, E., Tomasiewicz, H. G. & Erikson, R. L. (1984) Mol. Cell. Biol. **4**, 77–85
2. Gerke, V. & Weber, K. (1984) EMBO J. **3**, 227–233
3. Gerke, V. & Weber, K. (1985) J. Biol. Chem. **260**, 1688–1695
4. Gerke, V. & Weber, K. (1985) EMBO J. **4**, 2917–2920
5. Glenney, J. R., Jr., Tack, B. F. & Powell, M. A. (1987) J. Cell Biol. **104**, 503–511
6. Hexham, J. M., Totty, N. F., Waterfield, M. D. & Crumpton, M. J. (1986) Biochem. Biophys. Res. Commun. **134**, 248–254
7. Glenney, J. R., Jr. & Tack, B. F. (1985) Proc. Natl. Acad. Sci. U.S.A. **82**, 7884–7888
8. Glenney, J. R., Jr. (1986) J. Biol. Chem. **261**, 7247–7252
9. Kligman, D. & Hilt, D. C. (1988) Trends Biochem. Sci. **13**, 437–443
10. Baudier, J. & Gerard, D. (1983) Biochemistry **22**, 3360–3369
11. Kuznicki, J. & Filipek, A. (1987) Biochem. J. **247**, 663–667
12. Johnsson, N., Marriott, G. & Weber, K. (1988) EMBO J. **7**, 2435–2442
13. Zimmer, D. B. & van Eldik, L. (1986) J. Biol. Chem. **261**, 11424–11428
14. Filipek, A., Gerke, V., Weber, K. & Kuznicki, J. (1991) Eur. J. Biochem. **195**, 795–800
15. Osborn, M., Johnsson, N., Wehland, J. & Weber, K. (1988) Exp. Cell Res. **175**, 81–96
16. Zokas, L. & Glenney, J. R., Jr. (1987) J. Cell Biol. **105**, 2111–2121
17. Semich, R., Gerke, V., Robenek, H. & Weber, K. (1989) Eur. J. Cell Biol. **50**, 313–323

18. Powell, M. A. & Glenney, J. R., Jr. (1987) Biochem. J. **247**, 321–328
19. Drust, D. S. & Creutz, C. E. (1988) Nature (London) **331**, 88–91
20. Glenney, J. R., Jr., Boudreau, M., Galyean, R., Hunter, T. & Tack, B. F. (1986) J. Biol. Chem. **261**, 10485–10488
21. Johnsson, N., Vandekerckhove, J., Van Damme, J. & Weber, K. (1986) FEBS Lett. **198**, 361–364
22. Becker, T., Weber, K. & Johnsson, N. (1990) EMBO J. **9**, 4207–4213
23. Blumenthal, D. K., Takio, K., Edelman, A. M., Charbonneau, H., Titani, K., Walsh, K. A. & Krebs, E. G. (1985) Proc. Natl. Acad. Sci. U.S.A. **82**, 3187–3191
24. Lukas, T. J., Burgess, W. H., Prendergast, F. G., Lau, W. & Watterson, D. M. (1986) Biochemistry **25**, 1458–1464
25. O'Neill, K. T., Wolfe, H. R., Jr., Erikson-Virtanen, S. & DeGrado, W. F. (1987) Science **236**, 1454–1464
26. Strynadka, N. C. J. & James, M. N. G. (1990) Proteins **7**, 234–248
27. Amiguet, P., D'Eustachio, P., Kristensen, T., Wetsel, R. A., Saris, C. J. M., Hunter, T., Chaplin, D. D. & Tack, B. F. (1990) Biochemistry **29**, 1226–1232
28. Johnsson, N. & Weber, K. (1990) J. Biol. Chem. **265**, 14464–14468
29. Kube, E., Weber, K. & Gerke, V. (1991) Gene **102**, 255–259
30. Wasserman & Fullmer (1982) in Calcium and Cell Function (Cheng, W. Y., ed.), Vol. 2, pp. 175–216, Academic Press, Orlando, Florida
31. Huber, R., Roemisch, J. & Paques, E.-P. (1990) EMBO J. **9**, 3867–3784
32. Huber, R., Schneider, M., Mayr, I., Roemisch, J. & Paques, E.-P. (1990) FEBS Lett. **275**, 15–21
33. Thiel, C., Weber, K. & Gerke, V. (1991) FEBS Lett. **285**, 59–62
34. Isobe, T. & Okuyama, T. (1981) Eur. J. Biochem. **116**, 79–86
35. Isobe, T. & Okujama, T. (1978) Eur. J. Biochem. **89**, 379–388
36. Glenney, J. R., Jr., Kindy, M. S. & Zokas, L. (1989) J. Cell Biol. **108**, 569–578
37. Jackson-Grusby, L. L., Swiergiel, J. Q. & Linzer, D. I. H. (1987) Nucleic Acids Res. **15**, 6677–6690
38. Calabretta, B., Battini, R., Kaczmarek, L., de Riel, J. K. & Baserga, R. (1986) J. Biol. Chem. **261**, 12628–12632
39. Odink, K., Cerletti, N., Brueggen, J., Clerc, R. G., Taresay, L., Zwadlo, G., Gerhards, G., Schlegel, R. & Sorg, C. (1987) Nature (London) **330**, 80–82
40. Dorin, J. R., Novak, M., Hill, R. E., Brock, D. J. H., Secher, D. S. & van Heyningen, V. (1987) Nature (London) **326**, 614–617
41. Saris, C. J. M., Kristensen, T., D'Eustachio, P., Hicks, L. J., Noonan, D. J., Hunter, T. & Tack, B. F. (1987) J. Biol. Chem. **262**, 10663–10671
42. Masiakowski, P. & Shooter, E. M. (1988) Proc. Natl. Acad. Sci. U.S.A. **85**, 1277–1281

# Annexin II: involvement in exocytosis

**Robert D. Burgoyne**

The Physiological Laboratory, University of Liverpool, PO Box 147, Liverpool L69 3BX, U.K.

There is considerable interest in the mechanisms involved in exocytotic secretion. Many different hormones, neurotransmitters and a variety of other molecules are secreted from stored secretory vesicles or granules in response to a rise in cytosolic $Ca^{2+}$ concentration. The $Ca^{2+}$ signal has been studied in detail [1,2] but the nature of the receptor proteins for $Ca^{2+}$ that act in membrane fusion has still to be defined. Several $Ca^{2+}$-binding proteins have been identified in secretory cells and are suggested to be involved in $Ca^{2+}$-dependent exocytosis. Amongst these are members of the annexin family and in particular annexin II (calpactin). This review will consider the evidence suggesting a role for annexin II in $Ca^{2+}$-dependent exocytosis and will concentrate on data obtained from studies using adrenal chromaffin cells.

## Identification of annexins in adrenal chromaffin cells

The chromaffin cells of the adrenal medulla [3] have been used widely for the study of secretory mechanisms over the past 25 years because they provide access to large amounts of material including highly purified secretory granule fractions for biochemical study. The adrenal medulla was one of the first tissues shown to contain $Ca^{2+}$- and phospholipid-binding proteins now known to be annexins. These findings arose from attempts to isolate cytosolic proteins able to bind to chromaffin granules in a $Ca^{2+}$-dependent manner. The experiments were carried out in order to identify proteins that might be involved in $Ca^{2+}$-dependent exocytosis in chromaffin cells. Independent work from two separate laboratories detected a series of proteins in ‘cytosol’ fractions that bound reversibly either to immobilized chromaffin granules or chromaffin granule membranes in suspension [4–7]. Binding occurred at micromolar $Ca^{2+}$ levels and the proteins could be eluted from granule membranes by $Ca^{2+}$ chelation with EGTA. Several of these proteins have subsequently been clearly identified as annexins (annexins I, II, IV, V, VI and VII), although other proteins that also bind $Ca^{2+}$ include protein kinase C [8], phosphatidylinositol-specific phospho-

**Fig. 1. $Ca^{2+}$-dependent binding of cytosolic proteins (annexins) to chromaffin granules.**

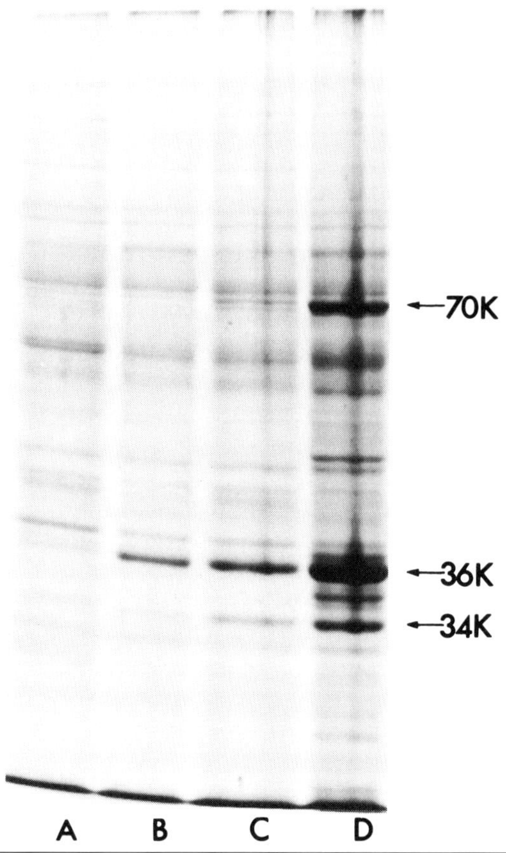


*Intact chromaffin granules were incubated with adrenal medullary cytosol at varying concentrations of free $Ca^{2+}$, and bound proteins were eluted with EGTA [5]. The PAGE tracks show proteins bound at 0.1 µM $Ca^{2+}$ (A), 1 µM $Ca^{2+}$ (B), 5 µM $Ca^{2+}$ (C) and 50 µM $Ca^{2+}$ (D). The 36 kDa polypeptide, binding at the lowest $Ca^{2+}$ concentration, is likely to be annexin II.*

lipase C [9] and the actin-binding protein caldesmon [10]. One of the binding proteins of molecular mass 36 kDa showed the lowest $Ca^{2+}$ requirement (Fig. 1); this protein is almost certainly annexin II.

The ability of the annexins to bind to chromaffin granules at micromolar $Ca^{2+}$ levels was taken to suggest that one or more of these proteins could be recruited to the granules on cell stimulation and then play some role in initiating or regulating secretory granule fusion with the plasma membrane [5,6]. An indication that the annexins could be involved in secretion came from the finding that two of them, annexin I [11] and annexin II [7], showed increased phosphorylation following cholinergic stimulation of chromaffin cells to induce catecholamine secretion.

## Secretory granule aggregation by the annexins

The first adrenal medullary annexin to be studied was annexin VII (synexin [12]), which was only discovered to be a member of this family after the other annexins had been cloned and sequenced. Annexin VII was identified as a protein that stimulated chromaffin granule aggregation (see chapter 9 by Pollard *et al.*). Subsequently, other annexins were found to aggregate chromaffin granules [7,13,14] with annexin II showing the highest $Ca^{2+}$ sensitivity [14]. The fact that annexin II was able to aggregate granules at physiologically relevant micromolar $Ca^{2+}$ concentrations (i.e. those that stimulate secretion from permeabilized cells) suggested that this annexin might be able to mediate membrane–membrane interactions leading to exocytosis. This idea was extended by the observation that granules aggregated by annexin II (and annexin VII) could be induced to fuse by addition of arachidonic acid [14]. Annexin II can exist as a monomer (heavy chain, p36) or a heterotetramer in association with its light chain ($p36_2\ p11_2$) ([15,16]; see chapter 6 by Weber). Drust and Creutz [14] found that while the heterotetramer was able to aggregate chromaffin granules, the monomer was ineffective. This result is surprising because, in the case of the aggregation of phospholipid vesicles, the annexin II monomer is almost as effective as the tetramer under certain conditions [17,18]. Only annexins I and II appear to be able to aggregate and stimulate fusion of phospholipid vesicles. In fact annexin VI inhibits spontaneous fusion in the presence of $Ca^{2+}$ [18]. The ability of annexin I to aggregate vesicles has been attributed to a domain stretching from amino acids 41–118, including part of the *N*-terminal tail and the first of the annexin repeats [19].

The ability of annexins I and II to aggregate and stimulate fusion of vesicles suggests that one or other of these proteins could be involved in exocytosis. Work on the localization of annexin II in chromaffin cells, as well as functional studies (see below), favour such a role for this protein.

## Localization of annexin II in chromaffin cells

Studies on chromaffin cells originally identified annexin II as a cytosolic protein that could bind to chromaffin granules. From a variety of studies it now appears that annexin II is present at only low levels in cytosol and is associated predominantly with the plasma membrane and, to a lesser extent, the chromaffin granule membrane. Immunofluorescence studies suggested that the majority of annexin II was present close to the plasma membrane [20]. Electron microscopy on fibroblasts showed that, in this cell type, annexin II is almost exclusively located on the inner surface of the plasma membrane [21]. This also appears to be the case for chromaffin cells [22]. Biochemical studies have, however, shown that a proportion of the annexin II is associated with chromaffin granules. The granule-bound annexin II acts as a peripheral membrane protein that can not be removed by $Ca^{2+}$ chelation [23]. Interestingly, the other annexins examined (I, IV and VI) were not associated with the granule membrane.

The specific association of annexin II with both the plasma and chromaffin granule membranes is against consistent with a role for this protein in exocytosis. Nakata *et al.* [22] have suggested that annexin II forms short filamentous cross-links that attach chromaffin granules to the plasma membrane following cell activation. The presence of annexin II on both membranes may promote self-assembly to form such cross-links. Cross-links were visualized between phospholipid vesicles aggregated by annexin II and in secreting chromaffin cells but it was not confirmed by immuno-labelling in intact cells that those cross-links were formed by annexin II.

## Functional studies on annexin II in chromaffin cells

The ability of annexin II to aggregate and fuse membrane vesicles and its localization in chromaffin cells made it a promising candidate for a component of the exocytotic machinery. Functional studies have shown, more clearly, that annexin II regulates $Ca^{2+}$-dependent exocytosis in chromaffin cells.

Permeabilized chromaffin cells provide a useful system for the study of the requirements for exocytosis as they allow direct activation of exocytosis by micromolar $Ca^{2+}$. Digitonin-permeabilized cells show a loss in secretory responsiveness over time as cytosolic proteins leak from the cells [24,25]. This secretory rundown can be exploited by using it as the basis for an assay of cytosolic or extracted proteins that are required for $Ca^{2+}$-dependent exocytosis.

Following digitonin-permeabilization, we found that incubation with purified annexin II maintained secretory responsiveness by slowing rundown [25,26]. Both annexin II heterotetramer and monomer are effective. The stimulation of secretion by annexin II shows the same $Ca^{2+}$-dependency as control secretion, occurs with a similar time course and is ATP-dependent, thus supporting the view that annexin II stimulates

**Fig. 2. Scheme showing possible role of annexin II in cross-linking secretory granule and plasma membranes leading to exocytosis following a rise in cytosolic $Ca^{2+}$ concentration.**

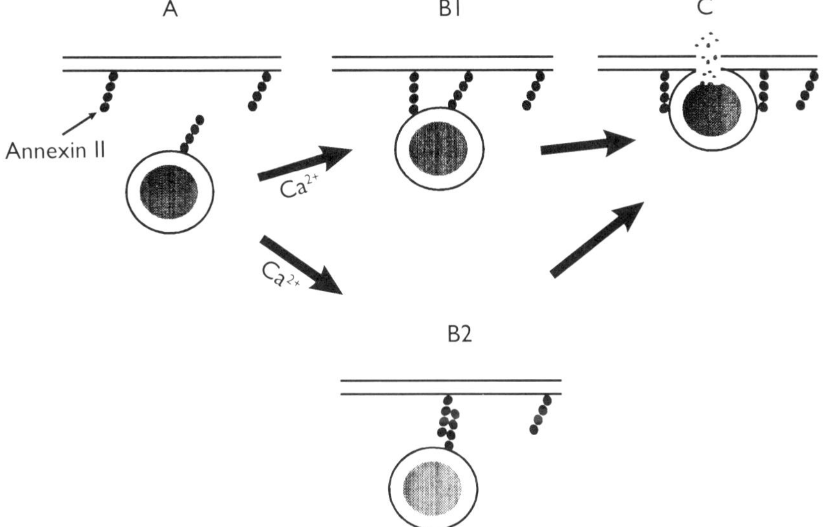


*Annexin II is shown associated with the plasma membrane and the secretory granule in resting cells (A), forming cross-links between plasma and granule membranes involving single (B 1) or multiple (B2) annexin II molecules, resulting in exocytosis (C).*

exocytosis and not simply granule lysis [25,26]. The stimulatory effect of annexin II required an intact *N*-terminus. Annexin II with the *N*-terminal tail removed by chymotrypsin treatment is inactive in the secretion assay [26]. Additional evidence that the *N*-terminal tail of annexin II regulates its function came from the finding that an affinity-purified antiserum raised against a synthetic *N*-terminal peptide [Ac-Cal (1–15)-$NH_2$] increased $Ca^{2+}$-dependent secretion when introduced into permeabilized chromaffin cells [27]. The exact mechanism leading to this stimulation of secretion by antibody binding to annexin II is unknown although one possibility is through an effect on phosphorylation which occurs in the *N*-terminal tail [28]. It was known from other studies that the *N*-terminal tail regulates the activity of the annexin II core domain [14,16,17,28]. A requirement for annexin II or one of the other annexins for exocytosis in chromaffin cells was also suggested by the finding that a 20-residue synthetic peptide, which was based on the conserved third domain of the annexins, partially inhibited $Ca^{2+}$-dependent secretion from permeabilized chromaffin cells [25].

The functional studies on annexin II in chromaffin cells [25–27] taken together with the morphological data on Nakata *et al.* [22] suggest the model in Fig. 2 in which annexin II forms initial cross-links between plasma and granule membranes following $Ca^{2+}$ influx. It is not known whether this would then be sufficient to lead to membrane fusion in the intact cells or whether other proteins are involved.

## Annexin II and exocytosis in other cell types

In the light of the information available on the role of annexin II in chromaffin cells, it is clearly of importance to determine whether annexin II is involved in exocytosis in other cell types. In the past, the distribution of annexin II has been examined using immunocytochemical approaches with antisera raised against the 'purified' protein. It is now clear that many of the original antisera cross-react with other annexins, thus making the early data unreliable. It is not completely certain, therefore, which cell types express annexin II. In the brain annexin II is present at low levels and does not appear to be concentrated in neurons [29]. In the anterior pituitary, it has recently been found to be bound to isolated secretory granules, whereas annexin I was not [30]; this situation is similar to that found in chromaffin granules. In PC12 cells, annexin II expression was markedly increased following nerve growth factor-induced differentiation, consistent with the protein being involved in growth or in a differentiated function such as neurotransmitter release [31]. Annexin II is an abundant protein in lung [32]; although its role in this tissue is unknown, one of the functions of the lung is the exocytotic secretion of surfactant. In ram sperms, both annexins I and II have been found to be associated with the acrosome which undergoes an exocytotic reaction [33].

Annexin II may also be involved in secretion in mammary epithelial cells. These cells differentiate during pregnancy and achieve their secretory phenotype after parturition. Using an affinity-purified antiserum against an *N*-terminal annexin II peptide, our group [34] found a striking appearance of annexin II on the secretory, apical membrane of mammary epithelial cells only after parturition. Thus, it would appear that annexin II becomes associated with the apical membrane late in differentiation as the cells start to secrete casein (Fig. 3). Our findings are in contrast with those of Lozano *et al.* [35] who claimed that annexin II expression in mammary cells was shut-off after differentiation; their finding that annexin II levels decreased may have resulted from the lack of specificity of the antiserum that they used for immunofluorescence. Further work is needed

**Fig. 3. Electron microscopical localization of annexin II in lactating mouse mammary epithelial cells.**

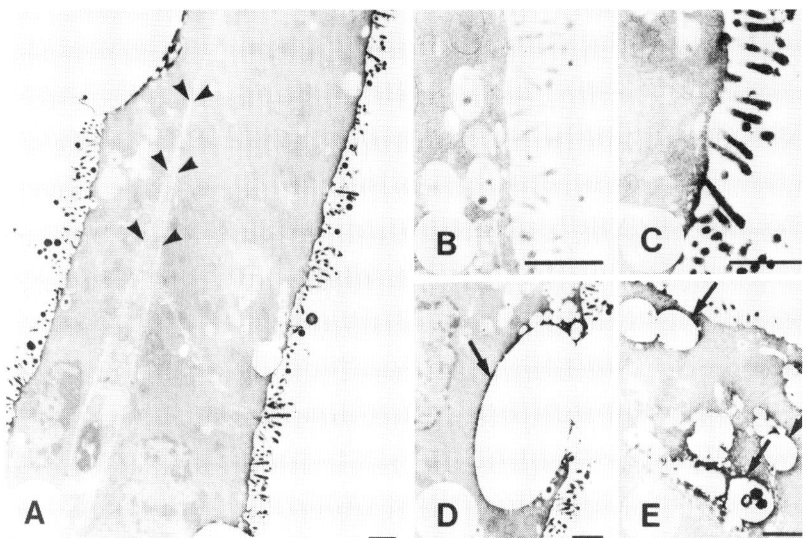


*Annexin II localization was determined using the immunoperoxidase technique with affinity-purified annexin II antiserum. The antiserum was raised against a synthetic peptide corresponding to the N-terminal 15 amino acids of annexin II [AC-Cal (1–15)-$NH_2$]. Annexin II was detected on the apical plasma membrane (A, C). An unstained section is shown for comparison in B. Immunostaining was also detected at sites of exocytosis (D) and on casein vesicles (E). Scale bars are 1 µm. Taken from [34] with permission.*

to determine whether annexin II does play a role in exocytosis in mammary cells, but it is interesting that these cells which had previously been thought to secrete constitutively have now been shown to possess a $Ca^{2+}$-dependent pathway for casein secretion [36].

As described above, indirect evidence has suggested that annexin II might be involved in exocytosis in several cell types. Functional investigations have yet to be carried out on cells other than the chromaffin cell to determine whether annexin II does indeed participate in exocytosis in these cell types. It is clear that annexin II must also have other functional roles since it is expressed in cells in which little exocytosis occurs. In addition, annexin II may not be a universal mediator of $Ca^{2+}$-dependent exocytosis as it has been found not to be expresed at high levels by all secretory cells.

## References


1. O'Sullivan, A. J., Cheek, T. R., Moreton, R. B., Berridge, M. J. & Burgoyne, R. D. (1989) EMBO J. **8**, 401–411
2. Cheek, T. R., Jackson, T. R., O'Sullivan, A. J., Moreton, R. B., Berridge, M. J. & Burgoyne, R. D. (1989) J. Cell Biol. **109**, 1219–1227
3. Burgoyne, R. D. (1991) Biochim. Biophys. Acta **1071**, 174–202
4. Creutz, C. E. (1981) Biochem. Biophys. Res. Commun. **103**, 1395–1400
5. Geisow, M. J. & Burgoyne, R. D. (1982) J. Neurochem. **38**, 1735–1741
6. Creutz, C. E., Dowling, L. G., Sando, J. J., Villar-Palasi, C., Whipple, J. H. & Zaks, W. J. (1983) J. Biol. Chem. **258**, 14664–14674
7. Creutz, C. E., Zaks, W. J., Hamman, H. C., Crane, S., Martin, W. H., Gould, K. L., Oddie, K. M. & Parsons, S. J. (1987) J. Biol. Chem. **262**, 1860–1868
8. Summers, T. A. & Creutz, C. E. (1985) J. Biol. Chem. **260**, 2437–2443
9. Creutz, C. E., Dowling, L. G., Kyger, E. M. & Franson, R. C. (1985) J. Biol. Chem. **260**, 7171–7173
10. Burgoyne, R. D., Cheek, T. R. & Norman, K. M. (1986) Nature (London) **319**, 68–70
11. Michener, M. L., Dawson, W. B. & Creutz, C. E. (1986) J. Biol. Chem. **216**, 6548–6555
12. Creutz, C. E., Pazoles, C. J. & Pollard, H. B. (1978) J. Biol. Chem. **253**, 2858–2866
13. Sudhof, T. C., Ebbecke, M., Walker, J. H., Fritsche, U. & Boustead, C. (1984) Biochemistry **23**, 1103–1109
14. Drust, D. S. & Creutz, C. E. (1988) Nature (London) **331**, 88–91
15. Gerke, V. & Weber, K. (1984) EMBO J. **3**, 227–233
16. Glenney, J. R., Boudreau, M., Galyean, R., Hunter, T. & Tack, B. (1986) J. Biol. Chem. **261**, 10485–10488
17. Powell, M. A. & Glenney, J. R. (1987) Biochem. J. **247**, 321–328
18. Blackwood, R. A. & Ernest, J. D. (1990) Biochem. J. **266**, 195–200
19. Ernest, J. D., Hoye, E., Blackwood, R. A. & Mok, T. L. (1991) J. Biol. Chem. **266**, 6670–6673
20. Burgoyne, R. D. & Cheek, T. R. (1987) Biosci. Rep. **7**, 281–288
21. Semich, R., Gerke, V., Robenek, H. & Weber, K. (1989) Eur. J. Cell. Biol. **50**, 313–323
22. Nakata, T., Sobue, K. & Hirokawa, N. (1990) J. Cell Biol. **110**, 13–25
23. Drust, D. E. & Creutz, C. E. (1991) J. Neurochem. **56**, 469–478
24. Sarafian, T., Aunis, D. & Bader, M. F. (1987) J. Biol. Chem. **262**, 16671–16676
25. Ali, S. M., Geisow, M. J. & Burgoyne, R. D. (1989) Nature (London) **340**, 313–315
26. Ali, S. M. & Burgoyne, R. D. (1990) Cell. Signal. **2**, 265–276
27. Burgoyne, R. D. & Morgan, A. (1990) Biochem. Soc. Trans. **18**, 1101–1104
28. Johnsson, N., Van, P. N., Soling, H. D. & Weber, K. (1986) EMBO J. **5**, 3455–3460
29. Burgoyne, R. D., Cambray-Deakin, M. A. & Norman, K. M. (1989) J. Mol. Neurosci. **1**, 47–54
30. Turgeon, J. L., Cooper, R. H. & Waring, D. W. (1990) Endocrinology (Baltimore) **128**, 96–102
31. Schlaepfer, D. D. & Haigler, H. T. (1990) J. Cell Biol. **111**, 229–238
32. Glenney, J. R., Tack, B. & Powell, M. A. (1987) J. Cell Biol. **104**, 503–511
33. Feinburg, J. M., Rainteau, D. P., Keatzel, M. A., Dacheux, J. L., Dedman, J. R. & Weinman, S. J. (1991) J. Histochem. Cytochem. **39**, 955–963
34. Handel, S. E., Rennison, M. E., Wilde, C. J. & Burgoyne, R. D. (1991) Cell Tissue Res. **264**, 549–554
35. Lozano, J. J., Silberstein, G. B., Hwang, S. I., Haindl, A. H. & Rocha, V. (1989) J. Cell Physiol. **138**, 503–510
36. Turner, M. D., Rennison, M. E., Handel, S. E., Wilde, C. J. & Burgoyne, R. D. (1992) J. Cell Biol. **117**, 269–278

8

# The annexins and exocytosis: some close approaches

**Carl E. Creutz, Christine Comera, Matthew Junker, Nicholas G. Kambouris, John R. Klein, Michael R. Nelson, Philip Rock, Sandra L. Snyder and Wei Wang**

Department of Pharmacology, University of Virginia, Charlottesville, VA 22908, U.S.A.

## Background: the annexins as potential mediators of exocytosis

The first annexin to be isolated was the protein synexin which was characterized as a mediator of calcium-dependent chromaffin granule aggregation [1]. The name synexin was derived from the Greek *synexis*, which means 'a meeting', because of the ability of this protein to promote intermembrane contacts. It was suggested that synexin could be a critical receptor for calcium in the process of exocytosis. A 32 kDa synexin homologue from bovine liver was named 'endonexin' in analogy with the term synexin [2], and this subsequently led to the introduction of the term 'annexin' for the family of structurally related calcium-dependent membrane-binding proteins [3]. At least eight members of this protein family have now been identified in a wide range of biological contexts [4]. Our research group has continued to investigate the hypothetical role of these proteins as mediators of membrane fusion in exocytosis.

More than 23 polypeptides bind to chromaffin granule membranes *in vitro* in a calcium-dependent manner [5–8]. These proteins have been called the 'chromobindins', in analogy with the established terms 'chromogranins', the secretory proteins in the chromaffin granule, and 'chromomembrins', the proteins in the granule membrane. These proteins may associate with the granule membrane in the stimulated chromaffin cell and direct the interactions of the vesicle with the cytoskeleton or the plasma membrane during exocytosis. The annexins comprise a major portion of the chromobindins: chromobindin 4 is annexin IV; chromobindins 5 and 7 appear to be variants of annexin V (see below); chromobindin 8 is annexin II; chromobindin 9 is annexin I; chromobindin 11 is annexin VII (synexin); and chromobindin 20 is annexin VI (p68) [8].

Most of the annexins exhibit the same 'bivalent' activity as synexin, that is, they can promote the aggregation of membranes. Annexin V may be an exception as we have not detected any ability of chromobindins 5 and 7 to aggregate chromaffin granules. In the case of chromaffin granules, after aggregation, membrane fusion is stimulated by the addition of free *cis*-unsaturated fatty acids, particularly arachidonic acid [9,10]. During exocytosis, this fusogenic agent might be provided through the activation of lipases by calcium and/or G-proteins. The bivalent activity of annexins generally is activated by higher concentrations of calcium than are required to promote binding of the annexin to a membrane. In fact, it has been found that the bivalent activity of the annexins requires higher levels of calcium than are likely to be available in the cytoplasm. However, the calpactin tetramer (annexin II) was found to promote membrane aggregation at levels of calcium (1–2 μM) that activate secretion from permeabilized chromaffin cells [10]. Thus, calpactin has been a particularly attractive candidate for a mediator of exocytosis in this cell (see chapter 7 by Burgoyne).

## Mechanism of the 'bivalent' action of the annexins

The recently described crystal structure of annexin V [13,15] suggests that the multiple membrane- and calcium-binding sites of the annexin core are all on one side of the planar molecule. It is not obvious, therefore, how the annexins could simultaneously bind to two membranes and promote aggregation. As there is some flexing of the molecule about a central cleft when calcium is bound [15], it is conceivable that the molecule might have enough flexibility to 'double over' and bind two membranes simultaneously. It seems more likely, however, that membrane-bound annexin molecules on two membranes might self-associate through their cytoplasmic faces to promote membrane aggregation. We proposed this as a mechanism of synexin action in promoting chromaffin granule aggregation when it was found that isolated synexin undergoes self-association with a calcium-dependence identical to its calcium-dependence for membrane aggregation but distinct from its calcium-dependence for membrane binding [16,17]. More recently we have demonstrated that synexin, endonexin and p68 can undergo self-association while promoting chromaffin granule aggregation [18]. To make these measurements, the annexins were labelled with fluorescein or rhodamine, and energy transfer between these fluorophores was interpreted as a measure of protein self-association. To verify that self-association was occurring between molecules on one membrane with molecules on another membrane, we bound fluorescein-labelled annexin to one population of chromaffin granules and rhodamine-labelled annexin to a separate population of granules. The two populations of granules were then mixed under conditions where the dissociation of the bound annexins from the membranes was negligible. The aggregation of the mixed granule population was then monitored by turbidity measurements, and the energy transfer indicative of intermembrane annexin self-association was measured simultaneously. As shown in Fig. 1, intermembrane self-association was seen at the lower levels of calcium that promote membrane aggregation, but was then lost at higher levels of calcium, although membrane aggregation still occurred. It appears, therefore, that at more physiological levels of calcium, annexin self-association may be required to drive membrane aggregation, but at supraphysiological calcium levels, the annexins might act monomerically to aggregate membranes. The self-association of annexins may depend on additional calcium-binding sites not previously recognized in the crystal structure of annexin V.

If self-association of membrane-bound annexin molecules is necessary in order to stimulate aggregation of membranes at physiological calcium concentrations, it is interesting then to consider the case of the calpactin tetramer. This complex molecule may already be, in a sense, in a self-associated form. The light chain is known to bind to the *N*-terminus of the heavy chain, and the structure of annexin V suggests the *N*-terminus of the heavy chain is likely to be on the cytoplasmic face of the membrane-bound heavy chain. A dimer of light chains may then hold the two heavy chains in proper orientation to bind two membranes simultaneously. The tetramer does indeed aggregate membranes at very low concentrations of calcium [10]. If this interpretation is correct, then one may expect the calcium-dependence of chromaffin granule aggregation by the tetramer to be identical to the calcium-dependence of chromaffin granule membrane binding by the heavy chain; an accurate measurement of the latter, for comparing with the published titration for granule aggregation, has yet to be determined [10].

Annexin VI (p68) has two ‘core’ domains as it has apparently evolved through gene duplication of a 30–40 kDa annexin. Thus, one may expect that, like the calpactin tetramer, it might aggregate membranes as a monomer and with high sensitivity to calcium. On the contrary, although p68 binds to membranes at 5–10 μM calcium, it aggregates membranes only at high concentrations of calcium (> 1 mM) and in fact inhibits membrane aggregation by other annexins [8,19,20]. Therefore, the membrane-binding faces of the two domains of p68 may be inappropriately

**Fig. 1. $Ca^{2+}$-dependence of intermembrane annexin self-association and membrane aggregation.**

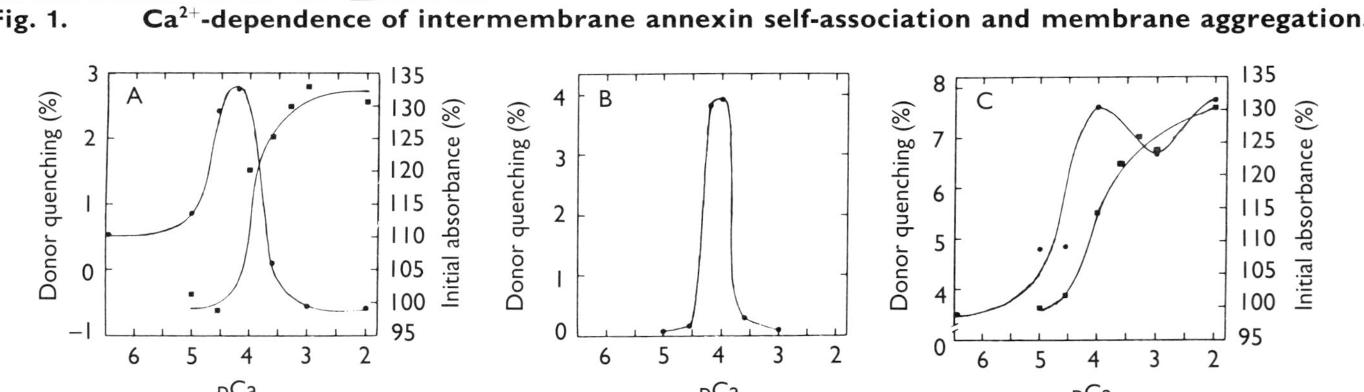


*For determination of intermembrane energy transfer two populations of chromaffin granule membranes (each 5 µg/ml) were incubated separately with equivalent amounts of fluorescein- or eosin-labelled endonexin (2.33 µg/ml each, A), synexin (2.02 µg/ml each, B) or p68 (6.8 µg/ml each, C). After the fluorescence emission intensity at 510 nm was recorded from the fluorescein-containing sample (containing 15–20-fold greater volume than the eosin-containing sample), $Ca^{2+}$ in varying concentrations was added to both samples. This produces a small decrease in fluorescein fluorescence as a result of self-quenching during self-association. When a stable fluorescence intensity was achieved (1–3 min) the eosin-labelled sample was introduced and the percent change in fluorescein quenching (●) expressed relative to the fluorescein fluorescence prior to $Ca^{2+}$ addition was determined as a measure of resonance energy transfer efficiency. The $Ca^{2+}$-dependence of chromaffin granule aggregation (■) was measured under identical total annexin and membrane concentrations. Reproduced from [18].*

oriented to aggregate membranes. High-resolution structural analysis of this protein will be of interest to verify this.

## Are the channel-forming properties of the annexins relevant to the mechanism of exocytosis?

Synexin and annexin V have been reported to form ion channels in lipid bilayers [11,12]. The annexin V crystal structure suggests that there is a hydrophilic channel through the centre of the protein, perpendicular to the plane of the membrane [13]. It is interesting to consider the relationship such a channel may have to the exocytotic pore that can be assessed by electrophysiological measurements of secretory cells [14]. The four homologous domains of annexin V are arrayed around the central hydrophilic channel [13]. The domains appear linked together by interaction of the *N*-terminal tail extending from the first domain to the fourth domain. If annexin molecules aggregate in the plane of the membrane, then perhaps the *N*-terminal tail of one molecule could interact with the fourth domain of a second molecule, resulting in eight domains surrounding an enlarged central channel. The process might then continue indefinitely to create a continuously expanding exocytotic pore. Some features of this model are not completely satisfying. For example, the annexin V molecule does not appear to integrate into the membrane, and therefore this expanding ring of polymerized annexins might only be able to stabilize one monolayer of the bilayer as the exocytotic pore is formed. However, it is indeed an interesting speculation that the annexins could be so multifunctional, in both promoting membrane contacts as well as regulating the controlled formation of a fusion pore between two membranes after they are brought into contact.

## Annexin mRNA levels in the chromaffin cell

We have recently conducted a survey of the mRNA levels for six annexins (annexins I, II, IV, V, VI and VII) in bovine chromaffin cell cultures (C. Comera and C. E. Creutz, unpublished work). We found that the levels of mRNA for these annexins was not altered by the action of secretogogues on these cells, after up to 1 h of stimulation. Thus, the secretory process does not appear to require the acute induction of specific annexins. It will be important in the future to determine if differentiation of a cell to become competent to carry out secretion, either during development or during manipulation of a model cell such as the pheochromocytoma PC12, might be associated with enhanced expression of particular annexins. Neurite outgrowth in PC12 has already been found to correlate with increased levels of annexin II protein [21].

## Two isotypes of annexin V (chromobindins 5 and 7) are expressed in the chromaffin cell

Two-dimensional tryptic peptide maps of chromobindins 5 and 7 were found previously to be indistinguishable [22], suggesting a close

**Fig. 2. Two mRNA species for annexin V are expressed in the bovine chromaffin cell.**

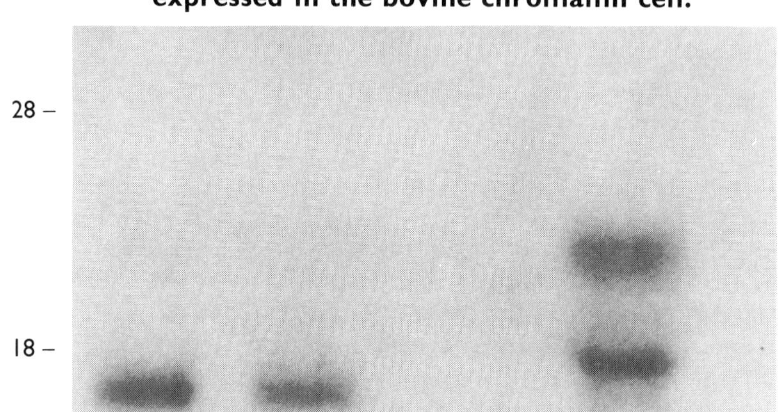


*The figure shows a Northern blot of mRNA from rat fibroblasts (R, Rat 1 cell line; L, LA29 cell line) and bovine chromaffin cells (C). The blot was probed with a full length human annexin V cDNA. The positions of the ribosomal 18S and 28S RNAs are marked.*

relationship between these two proteins. More recently, partial sequence analysis of peptides from two annexins isolated from bovine lung indicated that there may be two variants of annexin V in bovine tissues [23], although the sequence information was insufficient to judge whether one variant might be derived from the other by proteolysis. In our study of the annexin message levels in chromaffin cells by Northern blot analysis, we found that two annexin V messages are expressed at equal levels in chromaffin cells (Fig. 2). This contrasts with the situation in rat cells (Fig. 2) or human tissues [24] where only a single message is seen. It is likely that these two messages correspond to the two annexin V variants that have been partially sequenced, and to chromobindins 5 and 7.

## Genetic analysis of annexin functions: the search for yeast homologues

It seems likely that the annexin family of proteins has radiated to perform a variety of cellular functions. In that sense, the annexins may be similar to the other major class of calcium-binding proteins, the 'EF-hand' family which includes calmodulin, troponin C and the light chain of the calpactin tetramer, and which is responsible for a variety of non-overlapping cellular phenomena. If this is true, then not all annexins may be directly involved in exocytosis, even those described as chromobindins. On the other hand, more than one annexin might be involved in membrane trafficking, but at different steps in the secretory pathway, or through interaction with distinct secretory vesicle types such as chromaffin granules, synaptic vesicles or constitutive secretory vesicles.

An important potential approach for determining the functions of the annexins *in vivo* is through manipulation of the genes for these proteins. This is carried out most readily in a genetically tractable model such as yeast. Members of the annexin protein family have been identified in green plants [25] and in the slime mould *Dictyostelium discoideum* [26–28]. We have therefore sought to determine if annexins are also present in the yeast *Saccharomyces cerevisiae* and, if so, the effects on the physiology of the cell caused by annexin gene disruption. We have identified a family of yeast secretory vesicle-binding and phospholipid-binding proteins [29], and have cloned and deleted the genes for two of these proteins (N. G. Kambouris, D. Burke & C. E. Creutz, unpublished work). However, these two proteins proved not to be in the annexin class. One protein, the product of the gene we call *CAM1* (calcium and membrane-binding) is a homologue of the brine shrimp translation elongation factor (EF)1-γ [30]. The biochemical properties of this elongation factor subunit suggest it is responsible for membrane association and localization of the EF1 complex [31]. Our results suggest the localization might be calcium-dependent. EF1-γ forms a complex with EF1-β which in turn stimulates the exchange of GTP for GDP on EF1-α. Deletion of the *CAM1* gene did not result in an obvious phenotype, suggesting the role of EF1-γ in localization of protein synthetic machinery may not be important under laboratory conditions.

The second protein cloned is the product of the gene we call *YCP1*. It is a 47 kDa homologue of the cysteine protease bleomycin hydrolase [32]. Because this is a ubiquitous enzyme in mammalian cells it has been suggested to play a more fundamental role than metabolism of xenobiotics. However, disruption of the *YCP1* gene did not alter the basic growth properties of yeast cells.

As we are particularly interested in identifying true yeast annexin homologues, we are now attempting to clone the genes for a 45 kDa $Ca^{2+}$-dependent membrane-binding protein that reacts with an antiserum we prepared against an annexin consensus-sequence peptide and a 97 kDa protein that elutes from phenyl-Sepharose only in high concentrations of EGTA. In addition we are attempting to isolate yeast mutants that are dependent on mammalian annexin expression for survival. The genes altered in these mutants may encode yeast proteins that perform similar functions to the annexins.

The use of yeast as an experimental system raises the question of the relevance of the yeast secretory pathway to that of mammalian cells such as the chromaffin cell. The yeast pathway is responsible for the release of a polypeptide mating factor, hydrolytic enzymes and cell wall components. Although the expression of secretory proteins in yeast is closely controlled, once the proteins have entered the pathway their transit is currently viewed as uncontrolled, or 'constitutive'. In this regard the yeast secretory pathway may be a more appropriate model for constitutive secretion in mammalian cells, such as the secretion of plasma proteins from the liver, rather than regulated endocrine or exocrine secretion where release occurs from an accumulated pool of secretory vesicles. However, there is evidence that both pathways may involve common controlling elements. For example, small GTP-binding proteins, homologues of the SEC4 protein on yeast secretory vesicles, are also present on chromaffin granule membranes. It is our working hypothesis that regulated secretion may be fundamentally similar to constitutive secretion except that in the former case another level of control is exercised over the pathway so that secretory vesicles are held in check and then allowed to continue on the pathway only in response to a specific external stimulus. If this is the case, analysis of the constitutive pathway may provide valid insights into the organization of the regulated pathway.

## Effects of the expression of mammalian annexins in yeast secretory mutants

Aside from the search for endogenous yeast annexin homologues, we have found that the yeast cell may provide a useful assay system for assessing the function of annexins *in vivo* [33]. A number of yeast mutants have been isolated that have temperature-sensitive defects in the secretory pathway (the *sec* mutants [34]). As a test of the hypothesis that the annexins can influence membrane trafficking in cells, we expressed five different annexins in 13 yeast secretory mutants [33]. The annexins were human lipocortin, endonexin II, synexin, murine p68 and bovine endonexin. We focused on the 10 'late' *sec* mutants, *sec1*, *2*, *3*, *4*, *5*, *6*, *8*, *9*, *10* and *15*, the lipid-transfer mutant, *sec14*, and the 'early' mutants, *sec17* and *sec18*. To perform these experiments we constructed a novel derivative of expression plasmid YEp51 that has a convenient *Nco*I restriction site for inserting mammalian cDNAs, as shown in Fig. 3. Expression from this plasmid is controlled by the *GAL10* promoter so that growth on glucose-containing

**Fig. 3. Partial restriction map and schematic diagram of expression vector YEpC²*1*, a derivative of YEp5*1* [40].**

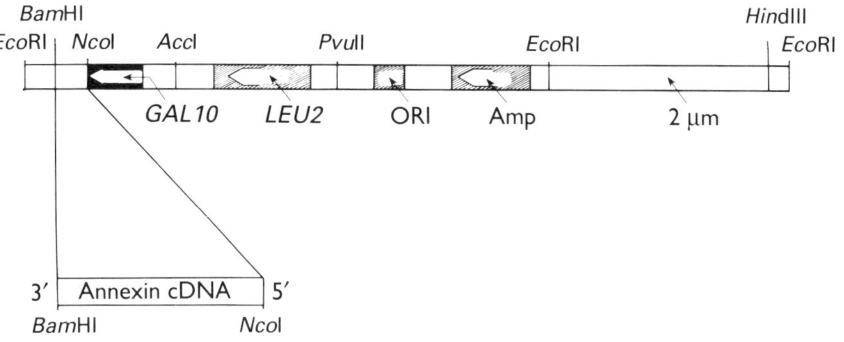


*Complementary DNAs may be inserted at the unique* NcoI *site downstream from the* GAL10 *promoter. Other features indicated are the* LEU2 *gene, the* Escherichia coli *origin of replication (ORI), the ampicillin-resistance gene (*Amp*) and the 2 μm circle sequences.*

medium, which represses annexin expression from the plasmid, can serve as a control for growth on galactose-containing medium. No effects of the annexins were apparent on the physiology of wild-type cells. None of the annexins fully complemented these *sec* mutants. However, synexin showed a strong dominant, negative interaction with *sec2*, *sec4* and *sec15*, and endonexin weakly supressed the growth defect of *sec2* mutants. In addition, synexin, lipocortin and p68 were found to accelerate the adaptation of glucose-grown cells to galactose-containing medium. Of the 65 constructs examined (13 *sec* mutants × 5 annexins) and compared with wild-type and null vector controls, these were the only interactions seen, thus emphasizing a high degree of specificity. The *SEC4* gene encodes a small GTP-binding protein that is present on the yeast secretory vesicle membrane [35]. The *SEC2* and *SEC15* gene products strongly interact with the *SEC4* gene product [36,37]. For example, overexpression of *SEC4* suppresses the *sec2* or *sec15* mutants. Therefore, our results suggest that synexin specifically interacts with a complex of proteins controlled by a regulatory GTP-binding protein that is directly responsible for fusion of the secretory vesicle with the plasma membrane.

The ability of the three annexins, synexin, lipocortin and p68, to accelerate the adaptation of yeast *sec2* mutants to galactose-containing medium when the cells were previously grown on glucose-containing medium [33], may be indirectly related to a membrane-fusion event. Part of the adaptation process involves the insertion of galactose-transport proteins into the plasma membrane [38]. In the *sec2* mutant this insertion process is partially defective [39]. We hypothesize that the annexins may serve to partially correct this defect by promoting the fusion of precursor vesicles containing the transporter molecules with the plasma membrane. The examination of this hypothesis may have important implications for similar processes in mammalian cell biology such as the insulin-responsive insertion of glucose transporters into muscle or fat cell plasma membranes.

To date, our analysis of the effects of annexin expression on the yeast secretory mutants has involved only measurements of cell growth rates [33]. It is now essential to directly measure secretion and the insertion of transport proteins in the plasma membranes of these cells to determine if the effects on growth can indeed be attributed to effects of the annexins on membrane trafficking in these cells.

## Conclusion

Although one of the important biological contexts within which the annexins were discovered is exocytosis, evidence defining their role in this process is incomplete. Their *in vitro* activities indicate that they are capable of promoting membrane fusion under physiologically relevant conditions. Just how they accomplish this is an important area of future investigation that will rely heavily on mutagenesis strategies guided in part by the rapidly emerging structural information on these proteins. Relating this activity to cellular function will rely on further work with permeabilized cells, genetic approaches and the development of specific pharmacological agents. The annexins promise many years of further close approaches in research, hopefully leading to the fusion of our multiple visions of their functions.

## Appendix: partial sequence of bovine annexin VI (by Christine Comera and Carl E. Creutz)

Bovine liver tissue is frequently used as a source of annexin VI for experimental studies, and the bovine chromaffin cell is an excellent model for analysis of the role of annexins in exocytosis. For these reasons knowledge of the sequence of bovine annexin VI is important. We isolated a near full length cDNA encoding bovine annexin VI by screening a bovine liver cDNA library [41] that was prepared in λ gt10. Oligonucleotides based on the published sequence of human annexin VI [42] were used as probes to screen the library. The partial cDNA sequences and

**Fig. 4. Partial sequence of bovine annexin VI.**

```
        V        S
  Q  E  I  C  Q  N  Y  K  S  L  Y  G  K  D  L  I  A  D  L    74
GGCAGGAGATCTGCCAGAACTACAAGTCCCTCTATGGCAAGGACCTCATTGCAGACTTG  222

K  Y  E  L  T  G  K  F  E  R  L  I  V  G  L  M  R  P  P  A    94
AAGTATGAGTTGACAGGGAAGTTTGAACGCCTGATTGTGGGTCTTATGAGGCCACCTGCC  282
   C
Y  A  D  A  K  E  I  K  D  A  I  S  G  I  G  T  D  E  K  C   114
TATGCTGATGCCAAAGAAATTAAAGATGCCATCTCGGGCATTGGGACCGATGAGAAGTGC  342
                                       M
L  I  E  I  L  A  S  R  T  N  E  Q  I  H  Q  L  V  A  A  Y   134
CTCATTGAGATCTTGGCTTCCCGGACCAACGAGCAGATCCACCAGCTGGTGGCAGCATAC  402
                  D                    I
K  D  A  Y  E  R  E  L  E  A  D  I  T  G  D  T  S  G  H  F   154
AAAGATGCCTACGAGCGAGAGCTGGAGGCTGACATCACTGGGGACACCTCTGGCCACTTC  462
Q
R  K  M  L  V  V  L  L  Q  G  T  R  E  E  D  D  V  V  S  E   174
CGGAAGATGCTCGTGGTCCTGCTGCAGGGTACCAGGGAGGAGGATGACGTAGTGAGCGAG  522
                  V
D  L  V  Q  Q  D  L  Q  D  L  Y  E  A  G  E  L  K  W  G  T   194
GACTTGGTGCAGCAGGATCTCCAGGACCTGTATGAGGCAGGGGAACTGAAATGGGGAACA  582

D  E  A  Q  F  I  Y  I  L  G  N  R  S  K  Q  H  L  R  L  V   214
GATGAAGCCCAGTTCATTTACATTTTGGGAAATCGCAGCAAGCAGCATCTCCGGTTGGTA  642

F  D  E  Y  L  K  T  T  G  K  P  I  E  A  S  I  R  G  E  L   234
TTTGATGAGTATCTGAAGACCACAGGGAAGCCGATTGAAGCCAGCATCCGAGGGGAGCTG  702
                                                      P
S  G  D  F  E  K  L  M  L  A  V  V  K  C  I  R  S  T  A  E   254
TCCGGGGACTTTGAGAAGCTGATGCTGGCTGTGGTGAAGTGTATCCGGAGCACCGCAGAG  762

Y  F  A  E  R  L  F  K  A  M  K  G  L  G  T  R  D  N  T  L   274
TATTTTGCTGAGAGGCTATTCAAGGCCATGAAGGGCCTGGGGACTCGGGACAACACCCTG  822

I  R  I  M  V  S  R  S  E  L  D  M  L  D  I  R  E  I  F  R   294
ATCCGCATCATGGTCTCCCGCAGTGAGCTGGACATGCTCGACATCCGGGAGATATTCCGG  882

T  K  Y  E  K  S  L  Y  S  M  I  K  N  D  T  S  G  E  Y  K   314
ACCAAGTATGAGAAGTCCCTGTACAGTATGATCAAGAACGATACCTCTGGCGAATACAAG  942
                  S
K  T  L  L  K  L  C  G  G  D  D  D  A  A  G  Q  F  F  P  E   334
AAGACTCTGCTGAAGCTGTGTGGGGGAGATGATGATGCTGCTGGCCAGTTCTTCCCGGAG 1002

A  A  Q  V  A  Y  Q  M  W  E  L  S  A  V  A  R  V  E  L  K   354
GCAGCGCAGGTGGCCTATCAGATGTGGGAACTTAGTGCAGTGGCCCGAGTAGAGCTGAAA 1062
                  N
G  T  V  R  P  A  G  D  F  N  P  D  A  D  A  K  A  L  R  K   374
GGAACCGTGCGCCCAGCTGGTGACTTCAACCCTGATGCAGACGCCAAAGCCTTGCGGAAA 1122

A  M  K  G  L  G  T  D  E  D  T  I  I  D  I  I  T  H  R  S   394
GCCATGAAGGGACTCGGGACTGACGAAGACACAATCATCGATATCATCACGCACCGCAGC 1182
   V
N  A  Q  R  Q  Q  I  R  Q     403
AACGCCCAGCGGCAGCAGATCCGGCAGA 1210
```

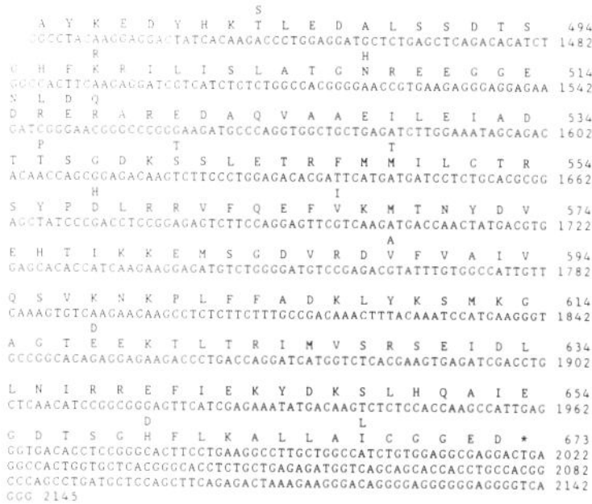

```
                          S
  A  Y  K  E  D  Y  H  K  T  L  E  D  A  L  S  S  D  T  S   494
GCCCTACAAGGAGGACTATCACAAGACCCTGGAGGATGCTCTGAGCTCAGACACATCT 1482
         R                             H
G  H  F  K  R  I  L  I  S  L  A  T  G  N  R  E  E  G  G  E   514
GGCCACTTCAAGAGGATCCTCATCTCTCTGGCCACGGGGAACCGTGAAGAGGGAGGAGAA 1542
N  L  D  Q
D  R  E  R  A  R  E  D  A  Q  V  A  A  E  I  L  E  I  A  D   534
GATCGGGAACGGGCCCGGGAAGATGCCCAGGTGGCTGCTGAGATCTTGGAAATAGCAGAC 1602
   P              T                       T
T  T  S  G  D  K  S  S  L  E  T  R  F  M  M  I  L  C  T  R   554
ACAACCAGCGGAGACAAGTCTTCCCTGGAGACACGATTCATGATGATCCTCTGCACGCGG 1662
         H                             I
S  Y  P  D  L  R  R  V  F  Q  E  F  V  K  M  T  N  Y  D  V   574
AGCTATCCCGACCTCCGGAGAGTCTTCCAGGAGTTCGTCAAGATGACCAACTATGACGTG 1722
                                          A
E  H  T  I  K  K  E  M  S  G  D  V  R  D  V  F  V  A  I  V   594
GAGCACACCATCAAGAAGGAGATGTCTGGGGATGTCCGAGACGTATTTGTGGCCATTGTT 1782

Q  S  V  K  N  K  P  L  F  F  A  D  K  L  Y  K  S  M  K  G   614
CAAAGTGTCAAGAACAAGCCTCTCTTCTTTGCCGACAAACTTTACAAATCCATCAAGGGT 1842
         D
A  G  T  E  E  K  T  L  T  R  I  M  V  S  R  S  E  I  D  L   634
GCCGGCACAGACGAGAAGACCCTGACCAGGATCATGGTCTCACGAAGTGAGATCGACCTG 1902

L  N  I  R  R  E  F  I  E  K  Y  D  K  S  L  H  Q  A  I  E   654
CTCAACATCCGGCGGGAGTTCATCGAGAAATATGACAAGTCTCTCCACCAAGCCATTGAG 1962
               D                       L
G  D  T  S  G  H  F  L  K  A  L  L  A  I  C  G  G  E  D  *   673
GGTGACACCTCCGGGCACTTCCTGAAGGCCTTGCTGGCCATCTGTGGAGGCGAGGACTGA 2022
GGCCACTGGTGCTCACGGGCACCTCTGCTGAGAGATGGTCAGCAGCACCACCTGCCACGG 2082
CCCAGCCTGATGCTCCAGCTTCAGAGACTAAAGAAGGGACAGGGGAGGGGGGAGGGGTCA 2142
GGG 2145
```

translated amino acid sequences of the clone are given in Fig. 4. The numbering is based on the numbering of the human sequence. The bovine cDNA as isolated begins at nucleotide 163. The region corresponding to nucleotides 1211–1424 was not sequenced. Amino acid substitutions present in the human sequence are indicated above the bovine amino acid sequence. The interpreted bovine amino acid sequence is 95% identical to the corresponding portions of the human sequence. The greatest area of divergence of the bovine and human sequences is at the beginning of the seventh repeat. The bovine sequence contains the six amino acids that correspond to an insertion in the human sequence (residues 525–530 in the seventh repeat [42]).

*The studies described here from the authors' laboratory were supported by research grants from the U.S. National Institutes of Health (DK33151 and CA40042). C.C. was the recipient of a fellowship from the Association pour la Recherche sur le Cancer (France).*

## References


1. Creutz, C. E., Pazoles, C. J. & Pollard, H. B. (1978) J. Biol. Chem. **253**, 2858–2866
2. Geisow, M. J., Fritsche, U., Hexham, J. M., Dash, B. & Johnson, T. (1986) Nature (London) **320**, 636–638
3. Geisow, M. J. & Walker, J. H. (1986) Trends Biochem. Sci. **11**, 120–123
4. Crumpton, M. J. & Dedman, J. R. (1990) Nature (London) **345**, 212
5. Creutz, C. E. (1981) Biochem. Biophys. Res. Commun. **103**, 1395–1400
6. Creutz, C. E., Dowling, L. G., Sando, J. J., Villar-Palasi, C., Whipple, J. H. & Zaks, W. J. (1983) J. Biol. Chem. **258**, 14664–14674
7. Geisow, M. J. & Burgoyne, R. D. (1982) J. Neurochem. **38**, 1735–1741
8. Creutz, C. E., Zaks, W. J., Hamman, H. C., Crane, S., Martin, W. H., Gould, K. L., Oddie, K. M. & Parsons, S. J. (1987) J. Biol. Chem. **262**, 1860–1868
9. Creutz, C. E. (1981) J. Cell Biol. **91**, 247–256
10. Drust, D. S. & Creutz, C. E. (1988) Nature (London) **331**, 88–91
11. Pollard, H. B. & Rojas, E. (1988) Proc. Natl. Acad. Sci. U.S.A. **85**, 2974–2978
12. Rojas, E., Pollard, H. B., Haigler, H. T., Parra, C. & Burns, A. L. (1990) J. Biol. Chem. **265**, 21207–21215
13. Huber, R., Romisch, J. & Paques, E. P. (1990) EMBO J. **9**, 3867–3874
14. Breckinridge, L. J. & Almers, W. (1987) Nature (London) **328**, 814–817
15. Huber, R., Schneider, M., Mayr, I., Romisch, J. & Paques, E. P. (1990) FEBS Lett. **275**, 15–21
16. Creutz, C. E., Pazoles, C. J. & Pollard, H. B. (1979) J. Biol. Chem. **254**, 553–558
17. Creutz, C. E. & Sterner, D. C. (1983) Biochem. Biophys. Res. Commun. **114**, 355–364
18. Creutz, C. E. & Zaks, W. J. (1991) Biochemistry **30**, 9607–9615
19. Zaks, W. J. & Creutz, C. E. (1990) Biochim. Biophys. Acta **1029**, 149–160
20. Pollard, H. B. & Scott, J. H. (1982) FEBS Lett. **150**, 201–206
21. Schlaepfer, D. D. & Haigler, H. T. (1990) J. Cell Biol. **111**, 229–238
22. Creutz, C. E. & Harrison, J. R. (1984) Nature (London) **308**, 208–210
23. Boustead, C. M., Walker, J. H. & Geisow, M. J. (1988) FEBS Lett. **233**, 233–238
24. Grundmann, U., Abel, K. J., Bohn, H., Lobermann, H., Lottspeich, F. & Kupper, H. (1988) Proc. Natl. Acad. Sci. U.S.A. **85**, 3708–3712
25. Smallwood, M., Keen, J. N. & Bowles, D. J. (1990) Biochem. J. **270**, 157–161
26. Gerke, V. (1991) J. Biol. Chem. **266**, 1809–1823
27. Doring, V., Schleicher, M. & Noegel, A. A. (1991) J. Biol. Chem. **266**, 17509–17515
28. Greenwood, M. & Tsang, A. (1991) Biochim. Biophys. Acta **1088**, 429–432
29. Creutz, C. E., Snyder, S. L. & Kambouris, N. G. (1991) Yeast **7**, 229–241
30. Maessen, G., Amons, R., Zeelan, J. & Moller, W. (1987) FEBS Lett. **223**, 181–186
31. Janssen, G. & Moller, W. (1988) Eur. J. Biochem. **171**, 119–129
32. Sebti, S. M., Mignano, J. E., Jani, J. P., Srimatkandada, S. & Lazo, J. S. (1990) Biochemistry **28**, 6544–6548
33. Creutz, C. E., Snyder, S. L., Kambouris, N. G., Hamman, H. C., Liu, W., Rock, P. & Klein, J. R. (1991) Abstracts of papers presented at the 1991 meeting on Yeast Cell Biology, Cold Spring Harbor Laboratory, p. 60, Cold Spring Harbor, New York
34. Novick, P., Field, C. & Schekman, R. (1980) Cell **21**, 205–215
35. Salminen, A. & Novick, P. J. (1987) Cell **49**, 527–538
36. Nair, J., Muller, H., Peterson, M. & Novick, P. J. (1990) J. Cell Biol. **110**, 1897–1909
37. Salminen, A. & Novick, P. J. (1989) J. Cell Biol. **109**, 1023–1036
38. Kew, O. M. & Douglas, H. C. (1976) J. Bacteriol. **125**, 33–41
39. Tschopp, J., Esmon, P. C. & Schekman, R. (1984) J. Bacteriol. **160**, 966–970
40. Broach, J. R., Li, Y. Y., Wu, L. C. C. & Jayaram, M. (1983) in Experimental Manipulation of Gene Expression (Inouye, M., ed.), pp. 83–117, Academic Press, New York
41. Hamman, H. C., Gaffey, L. C., Lynch, K. R. & Creutz, C. E. (1988) Biochem. Biophys. Res. Commun. **156**, 660–667
42. Crompton, M. R., Owens, R. J., Totty, N. F., Moss, S. E., Waterfield, M. D. & Crumpton, M. J. (1988) EMBO J. **7**, 21–27

# Synexin (annexin VII)

**Harvey B. Pollard, Eduardo Rojas, Natasha Merezhinskaya, Gemma A. J. Kuijpers, Meera Srivastava, Zhen-Yong Zhang-Keck, Anat Shirvan and A. Lee Burns**

Laboratory of Cell Biology and Genetics, National Institute of Diabetes, Digestive and Kidney Diseases, National Institutes of Health, Bethesda, MD 20892, U.S.A.

## Introduction

Synexin (annexin VII) was the first of the annexins to be discovered [1], but the seventh to be cloned and sequenced [2,3]. The problem leading to the discovery of this protein was how calcium might cause exocytotic fusion of secretory vesicles and plasma membranes, and the experimental question was whether a protein might exist which would transduce the calcium signal into such a fusion event. The cells in which we first found synexin were chromaffin cells from the adrenal medulla. The basis of the assay was an increase in the turbidity of a chromaffin granule suspension upon adding synexin and calcium. Subsequent work, to be described below, revealed that synexin is widely distributed in low abundance in most mammalian tissues, and is present throughout phylogeny, from slime mould to man. In the following sections we will describe our knowledge of the molecular biology of synexin and will summarize the changes in our understanding of how synexin interacts with target membranes. Finally, we will conclude with a critique of the specific relevance of synexin to exocytotic fusion.

## Molecular biology of synexin

The first hint that synexin might be a member of the annexin gene family was the detection of a set of partial amino acid sequences from the conserved *C*-terminal tetrad repeat domain of both human and bovine synexin [4]. Creutz *et al.* [5] also isolated peptides from bovine synexin, which eventually turned out to be from the unique *N*-terminal domain and so were not immediately useful for recognizing the annexin family designation. Efforts to isolate a synexin clone from bovine libraries were uniformly unsuccessful. However, our group [2] has isolated the human synexin cDNA gene and the details of this work are shown in Fig. 1.

The *C*-terminal tetrad repeat domain of human synexin proved to be about 50 % identical to other annexins [2], and has been found to share

**Fig. 1. Nucleotide and derived amino acid sequence of human synexin.**

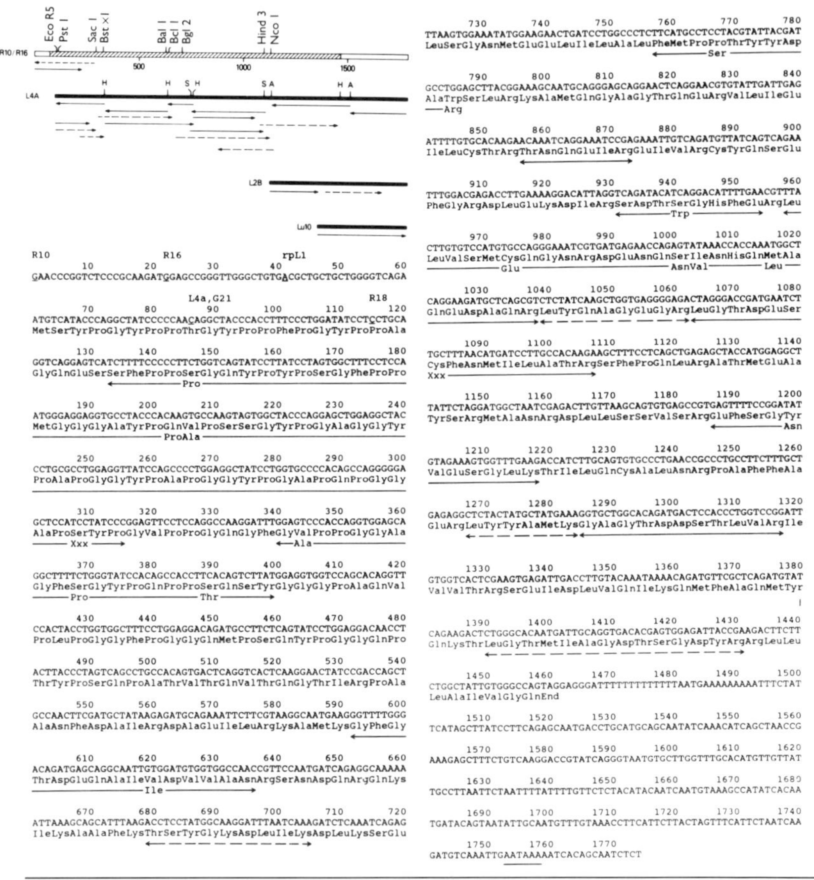


*Solid underlined arrows are identical to sequences from peptide sequence data of bovine synexin. Dashed underlined arrows are from human liver synexin peptides. Data are from [2].*

around 92% sequence similarity with mouse synexin (see below). By contrast, the unique *N*-terminal domain proved unusual even among the class of unique *N*-terminal domains of the other annexins. This domain is extraordinarily large (167 or 189 amino acids) and very hydrophobic. Indeed, of the entire length, only two amino acids are charged. The predicted secondary structure of human synexin was heavily β-pleated sheet and β-turn (see also [5] for further structural analysis). Furthermore, substantial parallels between human synexin and mouse synexin have also been noted (Z-Y. Zhang-Keck *et al.*, unpublished work).

Synexin gene sequences in a wide variety of mammals and yeast have also been detected on the basis of a Southern analysis using the human clone as a probe (A. Shirvan *et al.*, unpublished work). The synexin cDNA has also been cloned from the slime mould, *Dictyostelium discoideum* [7,8]. Two different but revealing approaches proved successful: one involved a nearly routine isolation of calcium-dependent membrane-binding proteins, and sequencing of a portion of the purified protein [7]; the other occurred as an incidental finding during a search for cyclic AMP-binding proteins in a *Dictyostelium* library [8]. In a recent retrospective study we found that the entire *N*-terminal domain of human synexin could, in fact, be shown to have significant sequence similarity to the equivalent domain of the cyclic AMP-binding protein (CABPI) from *Dictyostelium*. The localization of the region of the highest similarity, as well as the alignment of human synexin and CABPI are shown in Fig. 2. All amino acid substitutions in the indicated part of the *N*-terminal domain of human synexin and of CABPI are conserved. Further experimental work may reveal functional parallels to this otherwise compelling statistical relationship.

Detailed studies on the incidence and structure of the synexin mRNA in different tissues have revealed several polymorphisms, as well as a tissue-specific instance of expression of a cassette exon [9]. Northern analysis shows that the synexin message exists in liver and fibroblasts in two sizes; 2.0 and 2.4 kb. This distinction becomes important in light of the fact that the synexin gene occurs as only a single copy on chromosome 10, 10q21. The reason for the size heterogeneity is that the 3′-untranslated region occurs in either a shorter or longer sequence, based on selection of alternative poly(A) signals. Thus, regardless of the tissue samples, the larger mRNA has two poly(A) signals, whereas the shorter has only one. The underlying reason for such heterogeneity is not known, but seems specific for the human synexin gene. Indeed, studies on the mouse synexin cDNAs, presently in progress, show that observed size heterogeneity may occur by an entirely different mechanism.

A specific and predominant form of synexin occurs in brain, skeletal and cardiac muscle. In this form, 22 amino acids encoded by a cassette exon are added into the proximal one third of the *N*-terminal domain at position 145 [9]. Using the PCR technique it was shown that synexin mRNA containing the cassette exon could be detected in many regions of adult and fetal monkey brain, including frontal pole, dorsolateral

**Fig. 2. Comparison of the *N*-terminal domain of human synexin (synexin H) with the equivalent domain of the cyclic AMP-binding protein (CABPI) from *Dictyostelium*.**

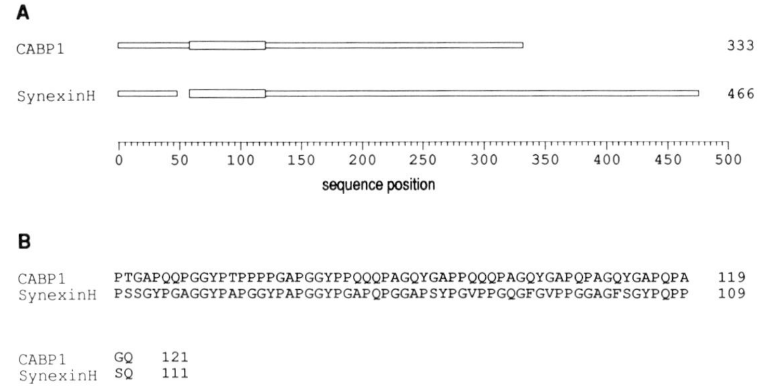


*A. Schematic representation of the region of homology between human synexin and CABPI. B. Alignment of the most conserved amino acid sequences of the N-terminal domain of human synexin and the comparable region of CABPI. Synexin and CABPI sequences were aligned and analysed using the Multiple Alignment Construction and Analysis Workbench (MACAW) program [46].*

prefrontal cortex, hoppocampus, caudate, visual cortex and somatosensory cortex. In these tissues the cassette-exon-containing message and gene product were the predominant species. By contrast, synexin lacking the cassette exon is the predominant (but not exclusive) species in liver, lung, kidney, spleen, fibroblasts and placenta. The basis of defining the mechanism of this polymorphism so specifically is our detection of the 66 bp exon sequence in genomic clones of synexin, flanked by introns with consensus acceptor and donor splice sites.

The tissue-specific distribution of the mRNAs with the cassette exon included is correlated with the predominant occurrence of a slightly higher-molecular-weight immunoreactive band in human skeletal muscle [9]. As expected, another polyvalent antisera generated against the peptide predicted by the cassette exon reacted only with synexin in human skeletal muscle. The antibody also reacted against recombinant human synexin expressed from cDNA with the cassette exon. Control Western blots were negative with samples of human lung, and with *Escherichia coli* extracts containing recombinant synexin without the cassette exon.

However, the nature of the additional sequences added to synexin by the cassette exon gives little insight into the possible functional significance of the change. The higher-molecular-weight synexin is otherwise active in the granule aggregation assay, and the bovine equivalent, isolated solely on the basis of size, is similarly active [10]. The presence of the exon does not affect the level of homology between synexin and CABPI because the additional cassette sequence is inserted next to the region of homology. The location of the insertion is also in the vicinity of the phosphorylation sites in some other annexins and, chemically, the insert introduces three charged amino acids into a domain otherwise remarkable for its hydrophobic character. The likely importance of these additional charges is the observation that human and monkey muscle synexin differ in this domain only by a serine replaced by a proline, and an aspartic acid replaced by a glutamic acid. The insert also contains a SYP sequence, which would appear to be a variation on the repetitive GYP motif found throughout the *N*-terminal domain.

The structure of the human synexin gene has also been determined in detail, and the information is presently in the process of being prepared for publication (A. Shivran *et al.*, unpublished work). Briefly, human synexin occurs as a single copy in chromosome 10 at position 10q21. The synexin gene is also large, encompassing in excess of 34 kb, and including 14 exons. Some of the introns are as large as 12 kb and the exon defining the cassette exon is placed exactly as expected in that part of the sequence coding for the unique *N*-terminal domain. Finally, splice junctions (5, 6, 7, 8, 9, 11) have been found to occur in similar sites to those defined previously for calpactin [11] and lipocortin [12] in the seqeunces corresponding to the first two repeats. However, all of the conserved splice sites, except for 11, are located either in conserved portions of the proximal end of the *N*-terminal domain or in the first two repeats. In a parallel study on the mouse synexin gene, also in the process of being prepared for publication (Z.-Y. Zhang-Keck *et al.*, unpublished work), the size and single-copy character seem to be preserved.

## Aggregation and fusion of chromaffin granules, ghosts and liposomes

The earliest studies on chromaffin granule aggregation revealed that synexin could drive the aggregation process only if calcium were also present [1]. However, the granule aggregates formed by synexin in the presence of calcium did not fuse. Nonetheless, the granules did not seem to be merely making a loose contact, since lowering the calcium concentration with EGTA after forming the aggregates did not cause dissociation of the complexes. Furthermore, dissociation could not be induced by high salt concentration. Indeed, only detergents and trypsin were found to be able to dissassemble the aggregates. Thus, a profound change seemed to be occurring at the site of membrane–membrane interaction, even though the aggregating granules did not actually fuse.

The mechanism by which synexin causes granules to aggregate appears to involve cross-linking of granules by synexin polymers [13]. Indeed, a very close parallel has been found between the processes of synexin polymerization and synexin-driven granule aggregation both in terms of specificity for calcium over other divalent cations, as well as for the apparent affinity constant (about 200 μM for bovine synexin). One difference was that the kinetics of calcium-driven synexin polymerization were much faster than granule aggregation occurring under virtually identical conditions of buffer, temperature, and concentration of synexin and calcium. Thus, it appears that the function of calcium is to cause synexin to polymerize, and that the now active polymeric protein species cross-links the granules.

Although the synexin-driven granule aggregation process stops short of complete fusion, the addition of small amounts of arachidonic acid can cause the granule aggregates to fuse into large vacuoles, easily visible by phase microscopy [14,15]. The products of arachidonic acid metabolism do not appear to be important for this process, in that other *cis*-unsaturated fatty acids are able substitutes for arachidonic acid. The fact that the vacuolar structures are formed has provided further impetus for the study of this process, as similar structures are seen in secreting chromaffin cells [16]. The latter are interpreted as evidence of compound or 'piggyback' exocytosis. In addition, there is evidence of concurrent activation of phospholipase $A_2$ during exocytosis from chromaffin cells, and this activity could account for the increased endogenous arachidonic acid produced during secretion [17,18].

However, the arachidonic acid fusion reaction cannot be taken as unambiguous support for a model of synexin-specific exocytosis. The reason is that arachidonic acid also causes fusion of granules aggregated by other annexins, such as calpactin I (annexin II, [19]) or lipocortin I (annexin I; G. Lee and H. B. Pollard, unpublished work). Thus, the molecular basis for the arrest of synexin activity on intact chromaffin granules in a pre-fusion state remains to be discovered.

By contrast, synexin is quite capable of causing the direct fusion of both lysed chromaffin granule ghosts as well as pure phospholipid liposomes. Stutzin [20] and Nir *et al.* [21] showed that chromaffin granule ghosts could be loaded with self-quenching concentrations of fluorescein isothiocyanate–dextran, and that synexin could drive the fusion of these ghosts with empty ghosts. Fusion could be followed easily by measuring the relief of self-quenching of the fluorophore upon dilution into the empty compartment. Inclusion of an anti-fluorescein isothiocyanate antibody, which quenches the fluorescence of fluorescein, provided a way to quench any signal caused by simple leakage or lysis of the loaded ghosts. Modelling of these results by a stochastic equation led to the conclusion that synexin not only promotes contact between fusing ghosts, but also drives the fusion event occurring after membrane contact (see also [38]).

Finally, quite some effort has been devoted to determining whether specific proteins as well as lipids might contribute binding sites for synexin on the membrane. Indeed, from studies with biological membranes alone it might not be possible to be certain about what factors on the membrane are responsible for synexin interactions. Like other annexins, however, calcium-activated bovine synexin has a specific affinity for acidic phospholipids, including phosphatidylserine, phosphatidylethanolamine, phosphatidic acid and phosphatidylinositol [22–25]. Liposomes prepared from phosphatidylserine and phosphatidylethanolamine will fuse directly in the presence of calcium and synexin [23], whereas phosphatidylinositol liposomes will merely aggregate [23]. Indeed, the cytoplasmic surface of the granules are enriched in the acidic phospholipids, and the granule membranes also contain an ATP-dependent flip-ase to ensure this orientation [26]. Thus, a role for a protein receptor for synexin on granules seems less compelling. However, the resistance to complete fusion by intact granules leaves unresolved the question of other modulating factors.

## Calcium channel activity of synexin and other annexins

Synexin is able to function as a calcium channel in both patch pipette [2,3,27] and planar lipid bilayer systems [28]. These data thus indicate that synexin not only binds to the membrane surface, but is also able to both penetrate and span the target membrane. The first indication that synexin could establish a more intimate association than mere adherence with membrane surfaces was the observation of capacitative currents in membranes exposed to calcium-activated bovine synexin [29]. Capacitative currents represent the movement of protein dipoles within the low dielectric interior of the membrane, and they are best known from analysis of sodium channels [30]. Later, calcium channels were observed with not only natural-abundance bovine synexin [27], but also natural-abundance [2] and recombinant [31] human synexin.

The first studies on human synexin channels showed that a remarkable parallel existed between the selectivity of the synexin channels for calcium and the profound specificity for calcium observed in the synexin-driven granule aggregation process. Further parallels were noted in terms of sensitivity to phenothiazine drugs (e.g., trifluoperazine and promethazine; see [32]), and insensitivity to both cadmium (1 mM) and anti-L-type channel-blocking drugs of the verapamil/nifedipine class. In addition, synexin channels were easily formed in bilayers of acidic phospholipids, but not pure phosphatidylcholine. In the patch-pipette system, the synexin channels were also found to be voltage-sensitive. The open time probability was greater when the pipette potential was positive, indicating that the synexin molecule is likely to insert into the membrane from the *cis* (bath) compartment in an asymmetric manner. This interpretation also seems consistent with the results of our study on the voltage-dependence of synexin-based capacitive current transients [29].

At least three other annexins express calcium channel activity in the patch pipette or planar lipid bilayer system. These include endonexin II (annexin V; [33]) and lipocortin I (annexin I; [28]), as well as calelectrin 67K (annexin VI; [28] and N. Arispe *et al.*, unpublished work). However, the other annexins have channel properties that clearly distinguish them from synexin. For example, endonexin II (annexin V) channels also conduct divalent cations such as barium, and monovalent ions such as $K^+$, $Na^+$, $Cs^+$ and $Li^+$. This is reminiscent of the selectivity properties of the ryanodine receptor calcium-release channel in mammalian muscle, and so need not be a significant basis for surprise. In fact, channels measured in the presence of either caesium or lithium proved easier to measure by automated techniques because calcium seemed to interact with a site within the endonexin II channel and partially block conductance ([33] and appendix therein). Lipocortin I (annexin I) also conducts other monovalent and divalent cations and has other kinetic properties that distinguish it from both synexin and endonexin II (annexin V). We have therefore concluded from these data that the common *C*-terminal tetrad repeat shared by these annexins is the most likely locus of the channel activity in these molecules. The substantial latitude observed in the specific cation selectivities of the different annexins may well be related to the fact that the identity of amino acids in this domain is only about 50%, although homology is around 90%.

From this perspective, the fact that synexin and other annexins form channels has proven valuable in our recent studies on the cystic fibrosis transmembrane regulator (CFTR). CFTR is a member of the ATP-binding cassette (ABC) gene family, and the domain associated with ATP binding, termed the nucleotide-binding fold (NBF), has been shown to share homology with annexins in the distal region of the first annexin repeat [34]. The conserved phenylalanine in all annexins in this region corresponds to the amino acid deleted in the vast majority of cystic fibrosis patients (i.e., the ΔF508 mutation). Our own analysis has revealed at least one additional domain of significant sequence similarity and several other regions of lesser similarity. In fact, of the approximately 41 entirely conserved amino acids in all presently reported mammalian annexins, 19 are also found in the NBF-1 and contiguous 'R' domains of CFTR. However, taking into account the entire sequence, the overall degree of statistical significance is much less.

As a test of whether this statistical parallel between the NBF and the annexins was also functional, we examined recombinant NBF-1 for ion channel activity in our planar lipid bilayer system. Because CFTR is thought to be an anion channel, we anticipated that the NBF domain might conduct anions. Indeed, NBF behaved as a chloride channel, with no evident ability to conduct a variety of cations [35]. These data thus not only serve as a control in terms of ion selectivity for the annexin channel data, but also introduce the concept that proteins from other gene families may employ specific domains and functions of the annexins for equivalent processes. Whether this is due to mere chance, convergent evolution, exon sharing or some common ancestry remains to be determined.

## Structure of the synexin channel and interpretation in terms of mechanism of membrane fusion

The physiological significance of the ion channel activity of synexin, or other annexins, remains to be discovered. However, the fact that channel activity occurs indicates that, once activated by calcium, the synexin molecule(s) are able to span the membrane. It has been our view that such a process may be important for our understanding of the mechanism of membrane fusion driven by synexin, and we have gained considerable insight into the possible details of this mechanism by a consideration of the recently reported crystal structure of endonexin II (annexin V; [36,37]). The reported structure provides specific information about the location of α-helical domains in the common *C*-terminal tetrad repeat when the protein is in an otherwise aqueous environment. On the basis of the common channel property when the protein is within a low dielectric bilayer, we have suggested that it is the *C*-terminal tetrad repeat domain that most likely interacts with the bilayer during fusion. In order to engage in this interaction, however, the structure of the domain may well be different from the water-soluble form depicted in the crystal structure. Indeed, Huber *et al.* [36] were faced with considerable problems trying to understand how endonexin II structure, as solved, could traverse a membrane.

Our approach has been to model the synexin (and thus other

**Fig. 3. A 'TIM barrel' model for the channel domain of synexin and other annexins.**

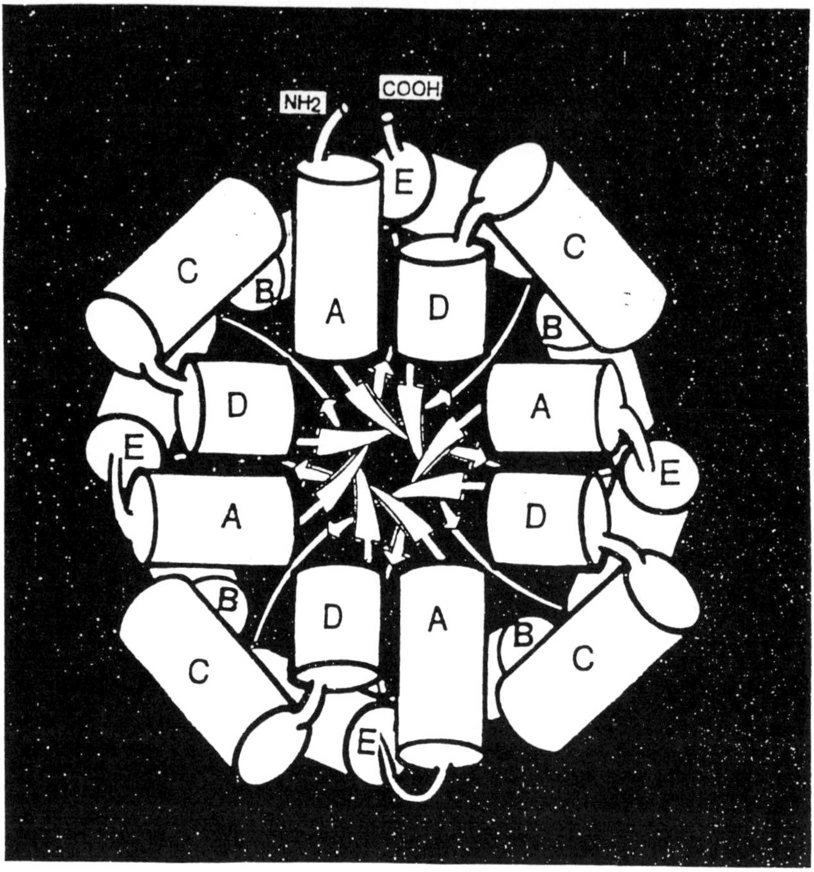

*The letters A–E correspond to α-helical domains, defined by crystallographic results and by calculation. The view shown is downwards into the shaft of the 'hollow screw' which traverses the membrane. The structure is formed from the four parts of the tetrad repeat. The β structures are formed from the amino acids between helices A and B, and between D and E, from each of the four repeats. The unique N-terminal domain is omitted from the structure, but would be adjoined to the region marked '$NH_2$'. Taken from [38].*

annexin) channel using the known α-helical domains, and information from modelling of other channels [38]. Specifically, on the basis of the eight-member anti-parallel β-barrel motif, which is likely to form the conducting pathway in potassium, calcium and sodium channels [39], we can predict that interhelical domains S1 (between helix A and B) and S2 (between helix D and E) can form an ion-conducting pathway. Each repeat can contribute two β-sheet structures, and thus a minimum of one synexin (or annexin) molecule can form a channel. Modification of the orientation of the five helices in each repeat allows the amphipathic structures to form an external hydrophobic sheath that permits ready interaction between synexin and membrane lipids. The model for formation of an annexin channel is shown in Fig. 3 [38]. We can summarize the model graphically as a 'hollow screw', penetrating the membrane. The view shown is down the 'hole' from the cytosolic side of the channel. The head of the screw faces the cytosol. Ions permeate the membrane through the polar internal hole, down a pathway delineated by the antiparallel β-pleated sheets. The exterior of the screw is hydrophobic in order to interact with lipids.

This view, in general, is central to the development of a model for membrane fusion which we have termed the 'hydrophobic bridge hypothesis'. This hypothesis is based on the previously discussed fact that synexin can form polymers when exposed to calcium. Our concept is that if a synexin polymer tries to form channel structures in two membranes simultaneously, the two membranes can be brought into contact for fusion. They are then crosslinked by a hydrophobic synexin polymer, and phospholipids from the adjoined membranes can then cross the hydrophobic bridge created by the synexin polymer, mix with each other, and fuse. We have shown how this could occur in schematic form, drawn approximately to scale (see [38] and a modification from [40] in Fig. 4). We have calculated that fusion by this process could be as rapid as 4 μs. This mechanism is based on the following: the known structural properties of synexin; the physically relevant properties of lipids in membranes, as derived from n.m.r. and molecular dynamics simulations; and the measured rates of synexin-driven membrane fusion. In principle, the mechanism could also be applied to other annexins if they had appropriate membrane-fusion properties. In our hands, some forms of annexin I can fuse acidic phospholipid liposomes, while annexins II, V and VI are inactive.

## Synexin and exocytosis: a critique

In the early 1980s several key problems confronted acceptance of the hypothesis that synexin might be the sought after mediator of exocytotic membrane fusion. One problem was that the $K_{\frac{1}{2}}$ for calcium, approximately 200 μM, seemed too high. For example, it was known that the calcium concentration in resting chromaffin cells was below 100 nM and that physiological stimulation caused the level to rise to as high as 500–1000 nM.

**Fig. 4. Hydrophobic bridge hypothesis for membrane fusion.**

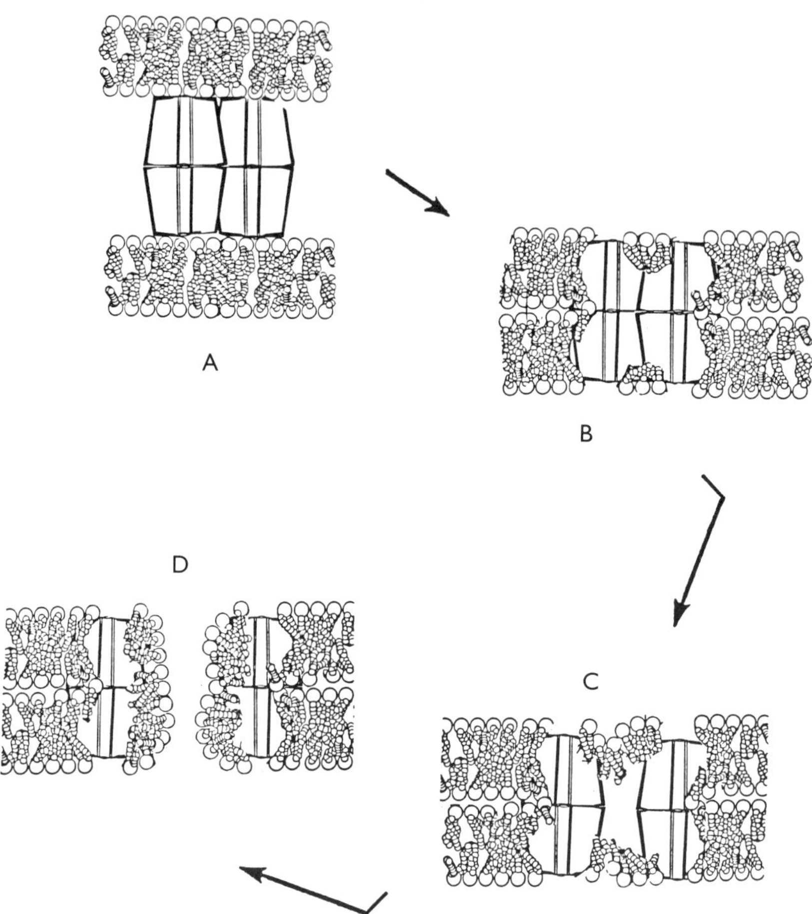


*(A) When exposed to calcium, synexin forms dimers and higher-order polymers, arranged in a side-by-side structure [47]. The polymers are highly hydrophobic and bind to phospholipid head groups on two adjoining membranes. The asymmetric structures of individual synexin monomers are based on a side-on view of the channel structure shown in Fig. 3. The N-terminal domain is not shown. (B) Phospholipids cross over the hydrophobic bridge provided by the synexin polymer. The ability of single synexin molecules to span a bilayer is based on the observed calcium channel activity. (C) The synexin polymer dissociates within the hydrophobic interior of the membrane, as the forces driving polymerization in the aqueous phase are now reduced. (D) Phospholipids migrate over the newly exposed synexin surfaces to complete the fusion process.*

How then could a protein with around 100-fold less sensitivity to calcium be a suitable detector and transducer in the secretion process? This seemed less important a problem for other members of the annexin gene family, such as calpactin I (annexin II), which has a substantially lower $K_{\frac{1}{2}}$ of about 2 μM [19]. With increasing experience and sophistication, however, it has become questionable whether the average concentration of free calcium in the cytoplasm might be an appropriate measure of the calcium concentration near the membrane surface, where the actual membrane contact and fusion processes actually occur. Indeed, Simon and Llinas [41] have shown by measurement and modelling that the concentration of calcium immediately beneath the open calcium channel in the plasma membrane could be as high as 1 mM. Thus, from this more recent perspective, an argument could be made that the apparent affinity for calcium by synexin might actually be in the appropriate range.

The selectivity of synexin for calcium was an additional problem, in that $Ba^{2+}$ and $Sr^{2+}$ were also believed to mimic calcium and promote exocytosis in chromaffin cells. How could synexin be responsible for exocytosis if it did not mediate action of other relevant divalent cations? This puzzle was resolved only recently when we learned that the kinetics of calcium-dependent and barium-dependent exocytosis could be followed simultaneously in the same chromaffin cell population [42]. Calcium was found to mediate a rapid and quickly terminated secretion process, while barium mediated a slow and sustained secretion process. The difference was also observed in permeabilized cells, leading to the conclusion that calcium and barium apparently promote secretion by separate mechanisms [42]. Parenthetically, we have searched for a soluble protein in bovine chromaffin cells that behaved like synexin in the presence of barium. This enterprise led to our isolation of a 36 kDa protein [43], now identified, by sequence of tryptic peptides from the unique *N*-terminal domain, as bovine lipocortin I (annexin I).

Another problem facing the synexin hypothesis was that the assay for activity, involving interaction among isolated chromaffin granules, seemed to be more of a reconstitution of compound exocytosis. Compound exocytosis occurs after simple exocytosis engineers fusion between a

---

*According to this model, the rate of fusion is determined by the diffusion constant of a lipid, and the maximum distance to be travelled (approximately 4 nm from each* trans *face). Fusion can thus be calculated to occur in around 4 μsec. As described in the text, this is consistent with data for synexin-driven chromaffin granule ghost fusion. The figure is a modification of the figure originally published in [38], as published in [40].*

chromaffin granule and the plasma membrane. Other granules then 'piggyback' on the initially fused granule and on each other. This is a dominant motif in chromaffin and other secreting endocrine cells. However, it has been shown that synexin can form a calcium-dependent complex with the inner aspect of the chromaffin cell plasma membrane [44]. Thus, interactions involving the plasma membrane, and leading to simple exocytosis, are not precluded in principle.

More recently, an argument in favour of a role for synexin in secretion has come from an immunoelectronmicroscopic analysis of changes in synexin localization in secreting chromaffin cells [45, 48]. Synexin, identified by the monoclonal antibody 10E7, is mainly cytosolic, but it is also specifically localized to chromaffin granules, plasma membrane and nucleoplasm. Stimulation of the cells with nicotine causes a reduction in the synexin antigen associated with the granules, and to a lesser extent with the other formed elements. Furthermore, these changes occur within the same time scale as the secretory process. A change of this sort is important because for a protein merely to be present in a secreting cell on membranes is hardly a compelling argument for its involvement in the secretory process. Thus, the observed change is consistent with a connection between synexin molecules and secretion. Furthermore, the result was observed in intact cells, and is thus immune from concerns about artifacts caused by leaky cells. However, it remains to be determined whether the connection between synexin changes and secretion is directly causal, or whether another related process is being recruited by the physiological stimulus.

## References


1. Creutz, C. E., Pazoles, C. J. & Pollard, H. B. (1978) J. Biol. Chem. **253**, 2858–2866
2. Burns, A. L., Magendzo, K., Shirvan, A., Srivastava, M., Alijani, M., Rojas, E. & Pollard, H. B. (1989) Proc. Natl. Acad. Sci. U.S.A. **86**, 3798–3802
3. Pollard, H. B., Burns, A. L. & Rojas, E. (1990) J. Membr. Biol. **117**, 101–112
4. Pollard, H. B., Burns, A. L. & Rojas, E. (1988) J. Exp. Med. **139**, 267–286
5. Creutz, C. E., Snyder, S. L., Huster, L. D., Beggerly, L. K. & Fox, J. W. (1988) Biochem. Biophys. Res. Commun. **152**, 1298–1303
6. Matsuchima, N., Creutz, C. E. & Kretsinger, R. H. (1990) Proteins Struct. Funct. Genet. **7**, 125–155
7. Gerke, V. (1991) J. Biol. Chem. **266**, 1697–1700
8. Greenwood, M. & Tsang, A. (1991) Biochim. Biophys. Acta **1088**, 429–432
9. Magendzo, K., Shirvan, A., Cultraro, C., Srivastava, M., Pollard, H. B. & Burns, A. L. (1991) J. Biol. Chem. **266**, 3228–3232
10. Dowling, L. G. & Creutz, C. E. (1985) Biochem. Biophys. Res. Commun. **132**, 382–389
11. Amiguet, P., D'Eustachio, P., Kristensen, T., Wetsel, R. A., Saris, C. J. M., Hunter, T., Chaplin, D. D. & Tack, B. F. (1990) Biochemistry **29**, 1226–1232
12. Kovacic, R. T., Tizard, R., Cate, R. L., Frey, A. Z., Wallner, B. P. (1991) Biochemistry **30**, 9015–9021
13. Creutz, C. E., Pazoles, C. J. & Pollard, H. B. (1979) J. Biol. Chem. **254**, 553–558
14. Creutz, C. E. (1981) J. Cell Biol. **91**, 247–256
15. Creutz, C. E. & Pollard, H. B. (1982) J. Auton. Nerv. Syst. **7**, 13–18

16. Ornberg, R. L., Duong, L. T. & Pollard, H. B. (1986) Cell Tissue Res. **245**, 547–553
17. Hotchkiss, A., Pollard, H. B., Scott, J. H. & Axelrod, J. (1981) Fed. Proc. Fed. Am. Soc. Exp. Biol. **40**, 256
18. Frye, R. A. & Holz, R. W. (1984) J. Neurochem. **43**, 146–150
19. Drust, D. S. & Creutz, C. E. (1988) Nature (London) **331**, 88–91
20. Stutzin, A. (1986) FEBS Lett. **197**, 274–280
21. Nir, S., Stutzin, A. & Pollard, H. B. (1987) Biochim. Biophys. Acta. **903**, 309–318
22. Hong, K., Duzgunes, N. & Papahadjopoulos, D. (1981) J. Biol. Chem. **256**, 3641–3644
23. Hong, K., Duzgunes, N., Ekert, R. & Papahadjopoulos, D. (1982) Proc. Natl. Acad. Sci. U.S.A. **79**, 4642–4644
24. Hong, K., Ekert, R., Bentz, J., Nir, S. & Papahadjopoulos, D. (1983) Biophys. J. **41**, 31a
25. Papahadjopoulos, D., Nir, S., Duzgunes, N. (1990) J. Bioenerg. **22**, 157–179
26. Zachowski, A., Henry, J-P. & Devaux, P. F. (1989) Nature (London) **340**, 75–76
27. Pollard, H. B. & Rojas, E. (1988) Proc. Natl. Acad. Sci. U.S.A. **85**, 2974–2978
28. Pollard, H. B., Guy, H. R., Arispe, N., de la Fuente, M., Lee, G., Rojas, E. M., Pollard, J. R., Srivastava, M., Zhang-Keck, Z-Y., Merezhinskaya, N., Caohuy, H., Burns, A. L. & Rojas, E. (1992) Biophys. J. **62**, 19–22
29. Rojas, E. & Pollard, H. B. (1987) FEBS Lett. **217**, 25–31
30. Rojas, E. (1976) Cold Spring Harb. Symp. Quant. Biol. **40**, 305–320
31. Burns, A. L., Magendzo, K., Srivastava, M., Rojas, E., Parra, C., de la Fuente, M., Cultraro, C., Shirvan, A., Vogel, T., Heldman, J., Caohuy, H., Tombaccini, D. & Pollard, H. B. (1990) Biochem. Soc. Trans. **18**, 1118–1121
32. Pollard, H. B., Scott, J. H. & Creutz, C. E. (1983) Biochem. Biophys. Res. Commun. **113**, 908–915
33. Rojas, E., Pollard, H. B., Haigler, H., Parra, C. & Burns, A. L. (1990) J. Biol. Chem. **265**, 21207–21215
34. Chap, H., Fauvel, J., Gassama-Diagne, A., Regab-Thomas, J. & Simon, M.-F. (1991) Medicin et Science **7**, 8–9
35. Arispe, N., Rojas, E., Hartman, J., Sorscher, E. & Pollard, H. B. (1992) Proc. Natl. Acad. Sci. U.S.A., in the press
36. Huber, R., Romisch, J. & Paques, E-P. (1990) EMBO. J. **9**, 3867–3874
37. Huber, R., Schneider, M., Mayr, I., Romisch, J. & Paques, E-P. (1991) FEBS Lett. **275**, 15–21
38. Pollard, H. B., Rojas, E., Pastor, R. W., Rojas, E. M., Guy, H. R. & Burns, A. L. (1991) Ann. N.Y. Acad. Sci. **635**, 328–351
39. Guy, H. R. & Conti, F. (1990) Trends Neurosci. **13**, 201–206
40. Pollard, H. B., Burns, A. L. & Rojas, E. (1992) Brain Res., in the press
41. Simon, S. M. & Llinas, R. R. (1985) Biophys. J. **48**, 485–498
42. Heldman, E., Levine, M., Raveh, L. & Pollard, H. B. (1989) J. Biol. Chem. **264**, 7914–7920
43. Lee, G., de la Fuente, M. & Pollard, H. B. (1991) Ann. N.Y. Acad. Sci. **635**, 477–479
44. Scott, J. H., Creutz, C. E., Pollard, H. B. & Ornberg, R. O. (1985) FEBS Lett. **180**, 17–23
45. Kuijpers, A. J. G., Lee, G. & Pollard, H. B. (1991) Ann. N.Y. Acad. Sci. **635**, 471–474
46. Schuler, G. D., Altschul, S. F. & Lipman, D. J. (1991) Proteins: Construct Anal **9**, 180–190
47. Creutz, C. E., Pazoles, C. J. & Pollard, H. B. (1979) J. Biol. Chem. **254**, 553–558
48. Kuijpers, A. J. G., Lee, G. & Pollard, H. B. (1992) Cell Tiss. Res., in the press

# Annexin V: crystal structure and its implications on function

**Robert Huber, Robert Berendes, Alexander Burger, Hartmut Luecke and Andrej Karshikov**

Max-Planck-Institut für Biochemie, D-8033 Martinsried, Federal Republic of Germany

## Introduction

Annexins constitute a family of cytosolic, water-soluble proteins, which bind to negatively charged phospholipids in a calcium-dependent manner (see [1–6] for reviews). They form a second family of calcium-binding proteins distinct from the 'EF-hand' family [7]. As its members were discovered independently, there are a variety of other names such as lipocortins, calpactins and endonexins (for a survey see [8]). The generic term 'annexin' was suggested by Geisow [9], in view of the similar biochemical and biophysical properties of these proteins.

Annexins can be found in a variety of cells and constitute a significant amount of cellular protein. They were discovered in evolutionary very distant organisms, including mammals (see [6] for a review), birds [10–12], amphibia [13], fish [14], *Drosophila* [15], *Dictyostelium* [16] and plants [17,18]. Ten members of this family have been characterized so far and their immunological [19,20] and biochemical properties (see [6] for a review) are closely related. Annexins IX and X have only been found in *Drosophila* [15].

Their primary structure is generally composed of four highly homologous repeats, with the exception of annexin VI, which has eight repeats (see [21] for a review). The *N*-termini are diverse and different in length for each annexin and can also be altered by alternative splicing [22]. Although a variety of *in vitro* properties have been described, out of which the ion channel activity of annexins V and VII is one of the most interesting [23–25], the *in vivo* role of the annexins still has to be established. Their widespread occurrence and the high degree of homology between all members of the annexin family (see [21] for a review) argues for important physiological roles.

In this chapter we will describe the three-dimensional structure of annexin V, which will help to explain the properties of the annexins down to the atomic level. We will also summarize the available electrophysiological data and propose possible structure-ion-channel relationships.

## The structure of annexin V

The first annexin to be structurally characterized was annexin VI (p68). Electron microscopy studies of two-dimensional crystals of lipid-monolayer-bound annexin VI revealed a trimeric arrangement of molecules resolved in six protein domains [26]. Three-dimensional crystals of annexin VI [26], annexin IV [27] and annexin V (human, rat) [28,29] have been obtained but no structure reported.

We studied annexin V from human placenta and presented the crystal structure and molecular model of annexin V for hexagonal and rhombohedral crystal forms at 2.3 Å and 2.0 Å resolution, respectively [30,31]. We showed that the protein comprising 320 amino acid residues is almost entirely α-helical (Fig. 1). Each of the four repeats of annexin V is folded into a compact domain of similar structure (Fig. 2). Five α-helices, wound into a right-handed superhelix, build up one domain (domain boundaries: tail, amino acid residues 5–16; domain I, 17–85; domain II, 88–157; connector, 160–167; domain III, 169–245; domain IV, 247–317). The axes of the four helices A, B, D, E are oriented approximately (anti-)parallel, whereas the connecting helix C lies almost flat. The helices are 7–15 amino acid residues long. They are rather regular α-helical. The domains I and II, and III and IV, respectively, are connected by short interhelical turns. An extended segment of about 10 residues forms the domain II and III connector. Similarly extended is the *N*-terminus up to residue 15.

The four domains are arranged in an almost planar, cyclic array. The domains II and III, and I and IV, respectively, give rise to tight modules with approximately twofold symmetry. A third local dyad (dyad C) relates modules (II, III) and (I, IV), so that all four domains have their molecular axes as defined by the axis of helices A, B, D and E similarly orientated. Dyad C marks the centre of the molecule and a very prominent hydrophilic pore, which we associate with the calcium-selective channel found in annexin VII and V [24,32]. The molecule has an overall flat, slightly curved shape with two faces, one convex and one concave. Five calcium-binding sites, all located at the convex face of the molecule, have been identified at high calcium concentrations.

**Fig. 1. Stereoviews of annexin V.**

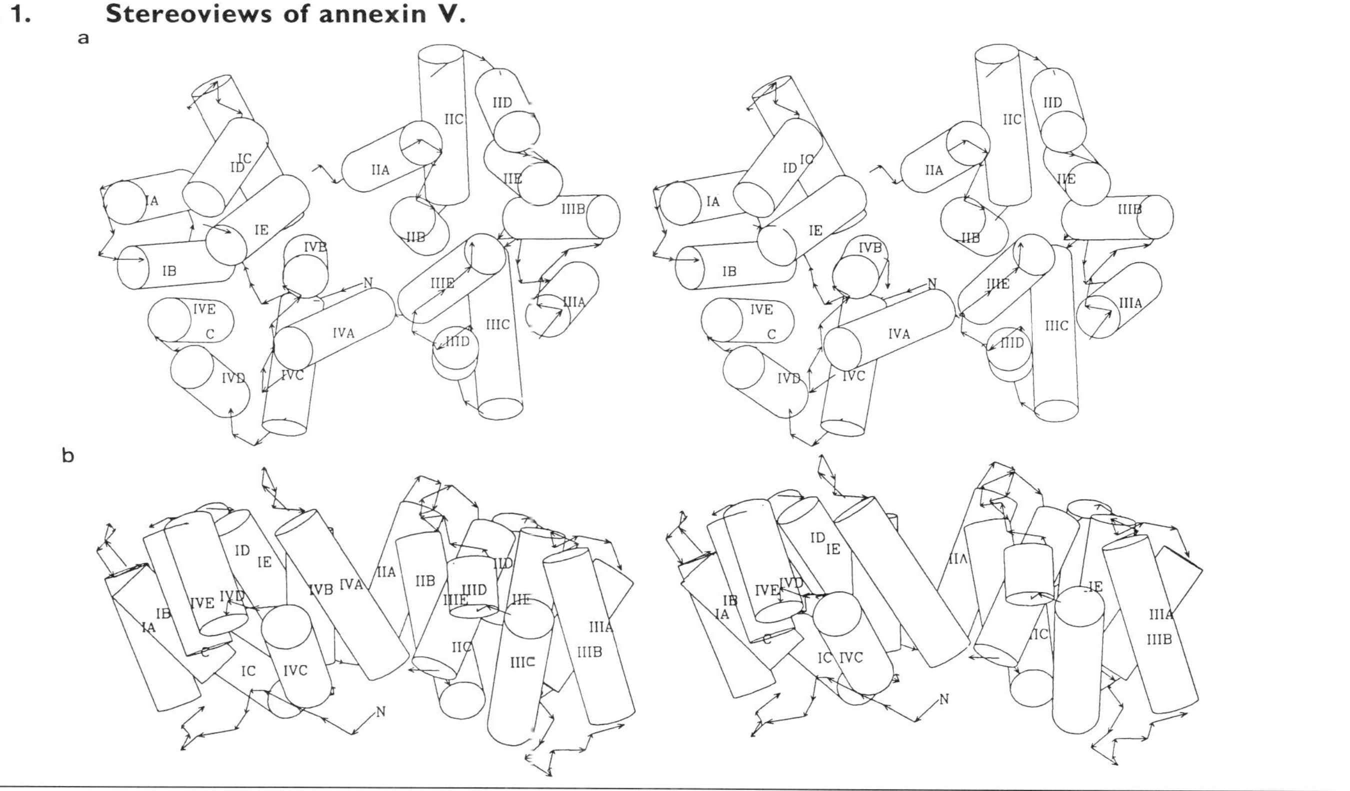


*The helical segments (A, B, C, D, E) are shown as cylinders and the four repeats are indicated by Roman numerals. Part (a) is the view onto the convex face from the membrane. The central pore appears widely open. Part (b) is the side view. The convex face on top is the calcium- and membrane-binding site.*

The projected structure of membrane-bound human annexin V has been resolved by electron-image analysis from negatively stained two-dimensional crystals [33]. These crystals were grown on planar lipid layers, composed of dioleoylphosphatidylserine and dioleoylphosphatidylcholine,

**Fig. 2. Stereoview of annexin V showing the $C^{\alpha}$ chain tracings of the four repeats superimposed.**

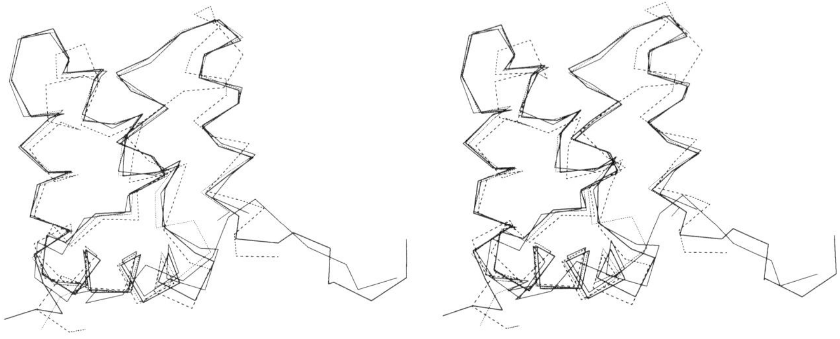

*Repeats I, II and IV match closely, but repeat III deviates, particularly at the calcium-binding site on top. Repeat III in annexin V does not bind calcium.*

by means of the lipid-layer crystallization technique [34]. Two-dimensional projection maps revealed triskelion-like motifs corresponding to annexin V molecules arranged in a hexagonal surface lattice with the symmetry of the plane group *p3*. Viewed perpendicular to the membrane plane, individual annexin molecules have dimensions and shapes very similar to the view seen along the central local dyad axis, relating modules (I, IV) and (II, III) [35]. The highly non-isometric and characteristic shape of the molecule allows no other superposition. The similarity extends beyond the level of the shapes of individual molecules and includes the packing of annexin both in the *p3* sub-lattice of the R3 crystals and in the *p3* two-dimensional lattice. Three- and two-dimensional crystals have closely related lattice parameters of 99.7 Å (R3) and 94 Å (*p3*) (1 Å = 0.1 nm) (Fig. 3).

## Family relationships

The members of the annexin family show a high amino acid sequence homology from the invariant F15 (annexin V numbering) on, which marks the onset of the globular structure. The most variable region of the

**Fig. 3.** **Electron microscopic projection map of two-dimensional crystals of annexin V superimposed with a trimer of annexin V molecules.**

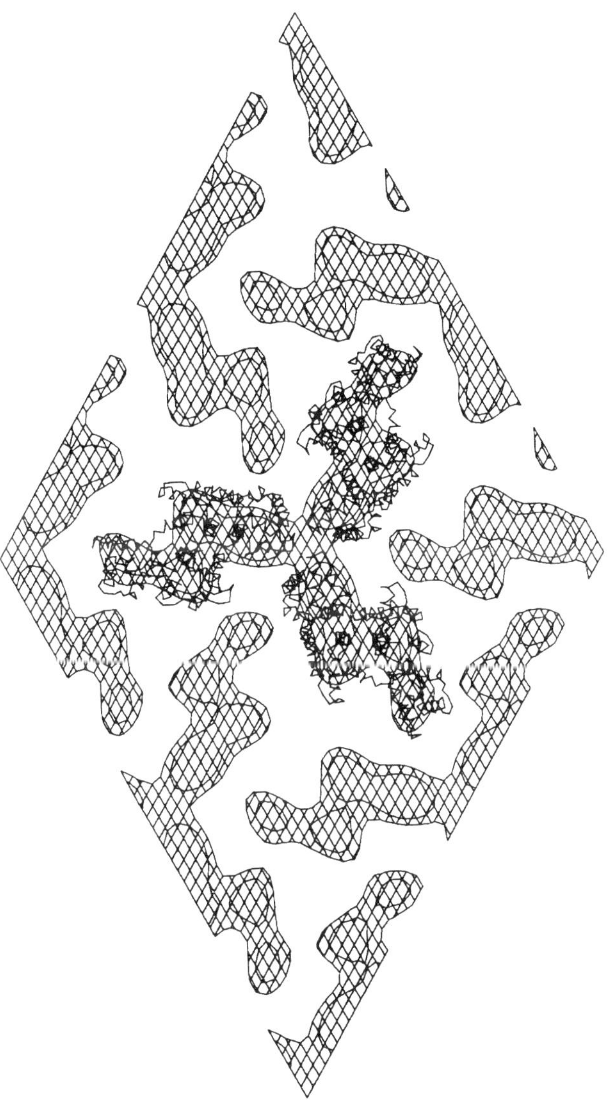

annexins is the *N*-terminal segment. It is short and has an extended conformation in annexin V. Repeats I and IV are linked non-covalently by the *N*-terminus, especially by residue 9, which is an isoleucine or valine in all annexins.

Interesting is the *N*-terminus of annexin VII (synexin). With 120 amino acid residues it is the longest in the annexin family and is characterized by a G/Y/P/Q-based sequence motif [24,36]. The function of the *N*-terminal tail of synexin is unknown, but may be related to binding of cytoskeletal proteins (see chapter 9 by Pollard *et al.*). More data are available concerning the *N*-terminal segment of annexin II (p36). It has 18 additional residues compared to annexin V and forms a well defined binding site for the p11 protein (see chapter 6 by Weber). An α-helical conformation of the *N*-terminus is crucial for the binding of p11, which leads to the formation of the heterotetramer $(p36,p11)_2$ [37,38].

In the globular part of the annexins about 40 residues are invariant and almost all replacements are conservative. The cores of the four repeats are formed by hydrophobic residues, except for Arg-50 in repeat I. The charge of this residue is balanced by three carbonyl oxygens, Leu-42, Leu-84 and Arg-45.

Profound differences exist between some of the calcium-binding loops of members of the annexin family. Whereas the loops for calcium 1 and 2, GXaaGT (102–105 and 261–265, respectively), in repeats II and IV are invariant in all annexins, a homologous loop in repeat I is found only in annexin V and some other annexins, but not in annexin I and II. A different sequence motif is seen in repeat III. Annexin V exhibits a tryptophan residue at position 187, which leads to a rather different course of the main chain traces in this part of the protein, with Trp-187 buried in the core of repeat III. In annexins I and II, Trp-187 is replaced by a lysine, which must adopt an alternative, probably exposed position, so that the loop may be able to bind calcium. Annexins I and II also have a substantial alteration of the calcium-binding motif in repeat I which reads V/A/TKGV 38 residues A/V and is probably not a calcium-binding site.

The presence of three calcium sites seems to be a common feature in annexins; two are invariable in repeats II and IV and one may be in either repeat I (all annexins except I and II) or III (annexins I and II). Annexin VI (p68) is the only annexin known so far to have eight repeats instead of four. On the basis of the homology of the two halves one can assume that each is structurally similar to the four-repeat annexins. Their relative arrangement is not constrained because of the long connector of 18 residues between repeats IV and V. Low-resolution images of the molecular shape of p68 from electron microscopy [26] suggest a lateral association such that all putative calcium sites in repeats I, II, IV, V, VI and VIII (according to the presence of the canonical ‘calcium’ sequence), may be similarly oriented.

## The calcium sites

Annexin I binds about four $Ca^{2+}$ ions with dissociation constants in the micromolar range [39,40]. The affinity for calcium is strongly increased in the presence of phospholipid [41]. Our group [31] crystallized annexin V in the presence of very high calcium concentrations to saturate the calcium sites even under high salt conditions, and we succeeded in localizing the binding sites in the crystal structure.

In the rhombohedral crystal form we resolved three almost fully occupied calcium ions (Ca1, Ca2, Ca3) bound to protruding polypeptide loops in homologous segments of repeats I, II and IV, and two lanthanum-binding sites in repeat I. The calcium ions and their coordinating peptide segments are not well defined in the hexagonal crystals, however. All calcium- and lanthanum-binding sites are located at the convex face of the molecule, as shown in Fig. 4.

Calcium prefers hepta-coordination by oxygen ligands located approximately at the vertices of a pentagonal bipyramid [41a]. Ca1 is hepta-coordinated to three carbonyl oxygens of Leu-100, Gly-102 and Gly-104, the bidentate carboxylate group of Asp-144, and two waters 481 and 478. The apices of the bipyramid are Leu-100 and water 478. Ca2 and Ca3 are similarly bound: Ca2 to Met-259, Gly-261, Gly-263, the bidentate Asp-303 carboxylate and water 483 (a second water is missing but there is weak density at the expected position); Ca3 to Met-28, Gly-30, Gly-32, the bidentate Glu-72, and two waters 403 and 706. The latter has weak density. The average calcium–oxygen distance is 2.42 Å. These three calcium-binding sites were first described by Huber *et al.* in Fig. 4 of [31] and have since been further refined [41b].

The structural relations between the calcium site of phospholipase $A_2$ [42] and the annexin V calcium sites Ca1 to Ca3 lead us to propose a model for the annexin–phospholipid interactions. A superposition of the phospholipase $A_2$ calcium loop with the Ca1 site as representative of all three calcium sites in annexin V is shown in Fig. 5 demonstrating the remarkable similarity between these two sites. In phospholipase $A_2$ the calcium is sequestered by a loop with sequence XXG(30)XG(32) and an aspartic acid at position 49 providing three carbonyl oxygens and two carboxylate proteinaceous oxygen ligands [42]. In a substrate analogue–inhibitor complex the two remaining coordination sites were found to be occupied by phosphoryl- and ester oxygens, respectively [43,44]. The

**Fig. 4. Stereoview of the $C^{\alpha}$ chain trace of annexin V and the three fully occupied calcium-binding sites (Ca1, Ca2, Ca3).**

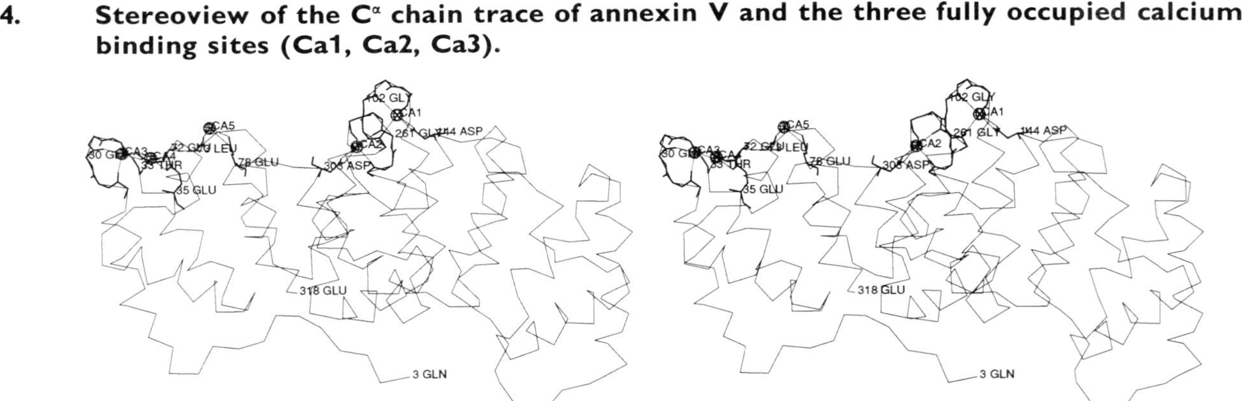


*Calcium-binding sites Ca1, Ca2 and Ca3 are in repeats II, IV and I, respectively. Ca4 and Ca5 are weakly occupied by calcium ions, but represent strong lanthanum-binding sites.*

loop and the aspartic acid match in structure (and sequence!) the calcium coordination with KGXGT(38 residues) E/D of Ca1 and similarly Ca2 and Ca3 in annexin V. No further structural similarities between phospholipase $A_2$ and annexin V exist. In addition, the general location of

**Fig. 5. Superposition in stereo of the calcium-binding sites of annexin V (Ca1: normal line) and phospholipase $A_2$ (bold line).**

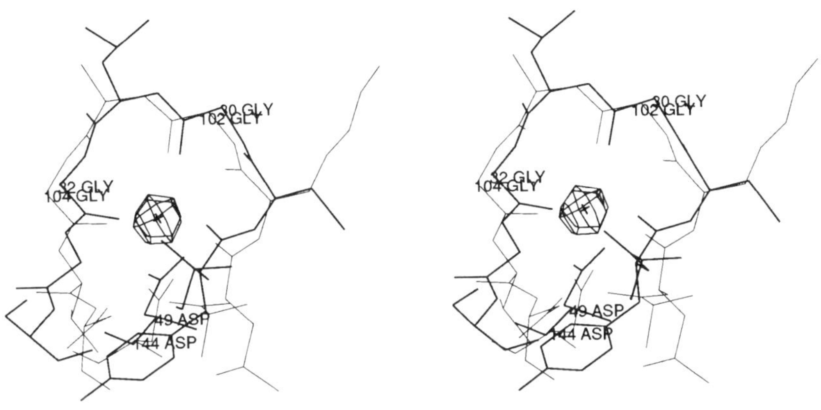


the conserved calcium loop is different, being very exposed in annexin V and in a depression in phospholipase $A_2$.

The calcium sites Ca4 and Ca5 in the rhombohedral form have calcium ions bound with low occupancies only. They were discovered in relation to their high affinity for lanthanum [31]. These sites are more open and offer only three proteinaceous oxygen ligands [Thr-33 and the bidentate carboxylate of Glu-35 (Ca4) and Lys-70, Leu-73, and the monodentate carboxylate of Glu-78 (Ca5)]. They may represent intermediate binding sites for $Ca^{2+}$ in transit through the channel.

Ca4 and Ca5 might also be phospholipid-binding sites. The observations that they are strong lanthanum-binding sites and that rare earth ions at low concentration block channel activity (R. Berendes, unpublished work) argue for a role in ion transport.

## Phosphorylation sites

Annexins I and II (lipocortin I and II) are substrates of the epidermal growth factor (EGF) kinase, the kinase encoded by the Rous Sarcoma virus oncogene ($pp60^{v\text{-}src}$) and protein kinase C. The sites of phosphorylation are located in the *N*-terminal segments [39,45–47]. Effects of phosphorylation on calcium- and phospholipid-binding are small and dissimilar for different annexins and phosphorylations [40,41]. Other properties have not been characterized. The *N*-terminal tail is close to the central pore and, although apparently not blocking it, may alter calcium flow when phosphorylated.

## Membrane–annexin interactions

A study of annexin V bound to phospholipid layers by electron microscopy and its comparison with our crystal structure did not indicate significant structural rearrangements [33,35] (Fig. 3). However, surface-loop or side-chain motions would not have been seen in the low-resolution electron micrographs.

As the phospholipid compositions and procedures of the electron microscopic preparations match those used for functional studies, the structures observed probably represent 'active' functional states. Because the annexin V molecule lacks rotational symmetry in the membrane plane, the face interacting with lipids could, in principle, be determined from projection maps, if the respective positions of the protein crystal, the lipid film and the support carbon film were known. This would be the case if protein two-dimensional crystals were associated with lipid monolayers. However, the orientation of annexin V could not be assigned unambiguously because a closer examination of the relationship between protein and lipid revealed the association of highly ordered two-dimensional crystals with complex assemblies such as bilayers (G. Mosser and A. Brisson, unpublished work).

Several other observations provide evidence that the main membrane-anchoring region of the annexin V molecule is located on the convex face. As mentioned, the five calcium-binding sites proposed to be directly involved in phospholipid binding are formed by protruding loops at the convex face. Their structural similarity to the calcium site in phospholipase $A_2$ [43] has been described above (Fig. 5). Several experiments have shown that changes in the *N*-terminal segments of annexins I and II (e.g. proteolytic cleavage) [40,48,49] or binding of small proteins and antibodies ([47]; N. Johnsson *et al.*, personal communication) have no profound effects on the membrane binding. Moreover, the spectral properties of the single tryptophan residue found close to the concave face in annexin I and close to the convex face in annexin V, respectively, are influenced by membrane binding only in the latter case [50]. Finally, the subtle effects of modifications in the *N*-terminal region, such as proteolysis and phosphorylation [39,41] could be related to allosteric interactions whose structural basis remains to be established. Concluding from these structural and functional data and taking into account the high amino acid sequence variability of the *N*-terminal region up to residue 15 (annexin V numbering) in the annexin family, the concave surface of the molecule can be excluded as the membrane-docking area.

The pronounced structural similarity of the annexin V calcium loops 1–3 with the calcium-phospholipid-binding site of pancreatic phospholipase $A_2$ suggests similar binding in annexin V through replacement of the calcium-bound waters by the phosphoryl and ester oxygens. There is a conserved basic site close to the putative phosphoryl-binding site, a possible determinant for acidic phospholipid binding [51,52]. This interaction offers an explanation for the strong cooperativity between calcium and phospholipid binding, implying a direct protein–phospholipid interaction in addition to the $Ca^{2+}$-mediated binding.

Assuming planar dimensions of 65 $Å^2$ for a phospholipid molecule [53], annexin V covers an area of about 26 phospholipids. Using tryptophan fluorescence measurements of annexin V bound to liposomes in the presence of 100 μM free $Ca^{2+}$, Meers *et al.* [54] estimated the amount of annexin V binding to phospholipid vesicles. By fitting theoretical binding curves to their experimental data they calculated a number of 59 phospholipids being affected by binding of a single annexin V molecule. A ratio of 42 affected phospholipids per annexin V molecule was evaluated on the basis of ellipsometric measurements [55]. Although these values describe the overall number of phospholipids influenced by annexin V membrane interactions, only a few acidic phospholipids are required for membrane attachment. An estimate of two to five molecules has been made [56], which is compatible with binding to three sites, Ca1 to Ca3.

No structural data are available for the bilayer membrane in the annexin V–membrane complex. Yet evidence for a profound membrane rearrangement is provided by experiments with a Langmuir trough. The surface pressure of phospholipid monolayers increases significantly upon incubation with annexin VI [26] and annexin V (R. Berendes *et al.*, unpublished work) in the presence of micromolar concentrations of $Ca^{2+}$. The generation of gated ion channels by annexins VII and V [24,25] (R. Berendes, unpublished work) further supports this conjecture.

A model for the annexin V–membrane interactions has been proposed on the basis of electrostatic calculations and our crystal structure of the annexin V molecule [56a]. The calculations performed included those for three negatively charged phospholipids. Their coordinates correspond to the three highest maxima of the electrostatic potential of the membrane–protein interface. The distribution of the electrostatic potential in the membrane space is strongly asymmetrical. In the region adjacent to the module (II, III) the electrostatic potential disappears close to the membrane–protein interface. The space adjacent to the module (I, IV), however, is characterized by a negative electrostatic potential. The protein–membrane interface covering almost the entire domain I and including the two $La^{3+}$-binding sites is characterized by a large patch of negative electrostatic potential equal to or stronger than $-0.2$ V. The potential becomes negligible at the distal leaflet of the membrane. Another smaller negative patch is seen on the protein–membrane surface between the domains II and III.

The electrostatic field calculated above reaches values sufficient for structural changes of the membrane similar to those in electroporation experiments [57–59]. Such a distortion of the membrane in the annexin V–membrane complex is important for the annexin V ion channel (see below). On the basis of the structural data and the model described above, we suggest that the specific calcium–phospholipid interactions and the non-specific effects of the annexin V electrostatic field lead to local membrane disorder and ion permeability.

## Annexins: Janus-faced proteins

Annexin V shares properties with integral membrane proteins. Our crystal structure in combination with the electron microscopy images shows that the central axis of pseudosymmetry of annexin V is normal to the membrane plane. Similarly, integral membrane proteins have a symmetry axis normal to the membrane plane (compare with the photosynthetic reaction centre [60]). This central axis, as in annexin V, defines a polar pore (compare with gap junction proteins [61] or the acetylcholine receptor; for reviews, see [62,63]). It is characterized by a very high helix content, in which most of the helix axes are oriented approximately parallel to the membrane normal (like in the photosynthetic reaction centre and in bacteriorhodopsin [64]).

In contrast to these similarities, the annexins show profound differences. They are water-soluble and have (most of) their polar residues projecting to the surface [30]. The few detailed structures of integral membrane proteins (the photosynthetic reaction centre [60] and bacteriorhodopsin [64]) demonstrate exposure of a hydrophobic surface to the surrounding lipids. The annexins might therefore, instead of integrating into the membrane, interact by a mechanism involving a profound membrane rearrangement. As a result the adjacent membrane bilayer or the proximal membrane leaflet would be disordered and thus made permeable to ions. The possible consequences of these rearrangements for the annexin ion channel activity are discussed below. Annexins display structural and functional features of soluble and integral membrane proteins. They face both phases (Janus-faced proteins) and mediate ion transport *in vitro*.

## Electrophysiology of annexins

Annexins interact with phospholipid membranes in a calcium-dependent manner. Some members form voltage-gated calcium-selective channels when associated with membranes, a property of integral membrane proteins.

Annexin VII (synexin) from bovine liver was the first annexin found to form voltage-gated $Ca^{2+}$-selective ion channels in bilayer membranes [23]. The *in vitro* channel activity of recombinant annexin VII was shown later [24]. Using the patch-clamp technique, annexin VII in the presence of 1 mM $Ca^{2+}$ was incorporated into phosphatidylserine bilayer membranes formed at the tip of the patch pipette. In single-channel measurements calcium ion currents in response to varying pipette potential were detected. $Ca^{2+}$ currents often occur in bursts and have a slope conductance of 26.5 pS in symmetrical 25 mM $Ca^{2+}$ solutions with a reversal potential of zero [23]. The open and closed time distributions could be fitted with single exponential functions. The resulting open and closed times are voltage-dependent. In ion-replacement studies it was shown that the annexin VII channel strongly prefers $Ca^{2+}$ over $Ba^{2+}$ or $Mg^{2+}$. Furthermore, annexin VII is blocked by the phenothiazine inhibitors trifluoperazine and promethazine in concentrations of about 50 μM. In contrast, only very high concentrations of $Cd^{2+}$ (> 10 mM), which is a potent blocker of other known voltage-gated $Ca^{2+}$ channels, block the annexin VII channel activity. Nifedipine inhibited annexin VII channel activity only at concentrations of 300 μM, 300 times more than the concentrations known to block non-inactivating $Ca^{2+}$ channels [23]. Thus, the annexin VII channel can easily be distinguished from other types of $Ca^{2+}$ channels.

Rojas *et al.* [25] established the $Ca^{2+}$ channel activity of annexin V in acidic phospholipid bilayer membranes. Human and recombinant annexin V form voltage-dependent cation channels when associated with bilayer membranes in the presence of $Ca^{2+}$. They prefer $Ca^{2+}$ over $Ba^{2+}$, $Sr^{2+}$ and $Mg^{2+}$, but also conduct $Li^+$, $Cs^+$, $Na^+$ and, to a lesser extent, $K^+$. In contrast to annexin VII, the analysis of the kinetic data obtained by single-channel recordings of annexin V showed that there are at least two open and two closed times [25, 56a]. The resulting two open and closed times are strongly voltage-dependent. While the open times increase in proportion to the voltage, the closed times decrease. Annexin V channel activity is not inhibited by millimolar concentrations of $Cd^{2+}$ or 50 μM nifedipine. $La^{3+}$ blockage of the annexin V channel activity at 600 μM was reported by Rojas *et al.* [25]. We found inhibition at $La^{3+}$ concentrations as low as 100 μM (R. Berendes, unpublished work). Therefore, the properties of the annexin V ion channel are more similar to annexin VII than to other known $Ca^{2+}$ channels.

Because the annexins are quite distinct from integral membrane proteins, the way in which annexin V and VII interact with the membrane and provide an ion pathway was speculative for some years. Our high-resolution crystal structure of annexin [30,31,41b] in combination with low-resolution electron-diffraction image analysis of phospholipid layer-bound annexin V [35] made it possible to propose a model for the annexin–membrane interaction, the ion channel formation and the ion conduction pathway.

The phospholipid bilayer membrane attached to annexin V has not been structurally characterized. The generation of gated ion channels by annexins VII and V provides evidence for a profound membrane rearrangement, however. We propose that the proximal and possibly also distal membrane leaflet may be disordered leading to lateral expansion and (non-selective) ion permeability, and suggest that the ion selectivity and gating properties of annexin V reside in the protein [30,31,41b]. Disorder may be caused by specific calcium–phospholipid interactions in a geometry dictated by the protein structure and by the electrostatic potential of the protein. It is known that external electric fields can invoke membrane pore formation (electroporation) under certain conditions [59]. Electrostatic potential calculations on annexin V show that the protein exerts a strong electric field in the region of the protein–membrane interface, and could thus contribute to membrane rearrangement through the electroporation effect [56a].

If, as we suggest, calcium–phosphoryl-group interaction and the protein's electrostatic field lead to local disorder, then the membrane area of annexin V associated with Ca1 to Ca3 may become ion-permeable. This area encompasses Ca4 and Ca5, which are strong lanthanum- and weak calcium-binding sites. The blockage of annexin V channel activity at micromolar concentrations of lanthanum mentioned above suggests these sites as candidates for intermediate binding sites for calcium ion conduction.

Direct experimental evidence for the ion-conduction pathway is not available. It may be contained in a single molecule or in an oligomer. Annexin V trimers are found in the R3 crystal form as well as in the annexin V phospholipid monolayer [35]. However, several observations indicate that the central molecular axis of annexin V is the ion-conduction pore: the invariant calcium sites 1 and 2, symmetrically arranged around the molecular axis mediate an intimate contact with the membrane, which is an important requirement for channel formation; the central pore is highly hydrated and almost all residues forming it are conserved; half of all buried charged residues participate in formation of the pore (Fig. 6); and more than half of all buried waters are localized in the pore and were found conserved in both crystal forms (Fig. 7). It is therefore highly probable that the chain of waters from Ca5 on through the central pore marks the ion conduction pathway [41b].

**Fig. 6. Stereoview from the membrane phase of the residues invariant in all annexins projected onto the $C^{\alpha}$ chain tracing.**

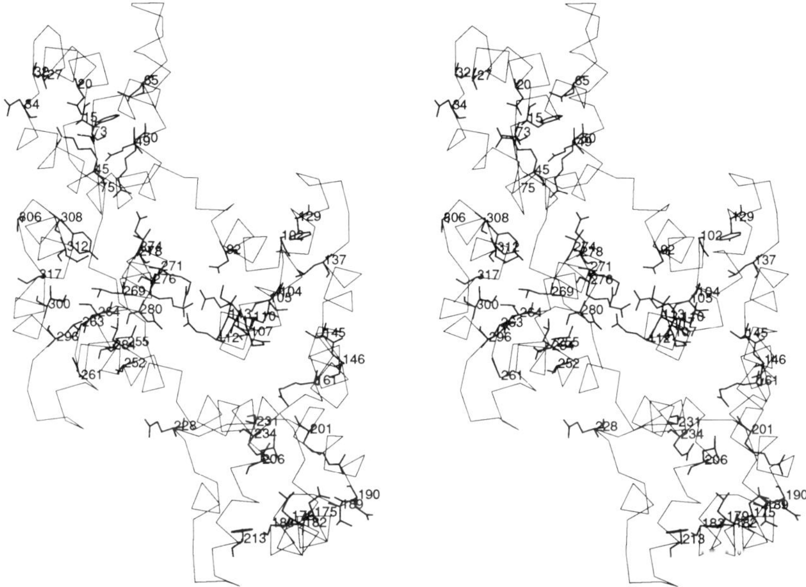

*The polar residues coating the central channel are strongly conserved.*

Other biological ion channels have not been characterized by their three-dimensional structures. Amino acid sequences and mutational and functional data lead one to expect that the pore of the potassium channel is associated with an extended rather than helical segment of hydrophobic nature and arranged as an eight-stranded β-barrel within the tetrameric aggregate ([66–68] and for reviews see [69,70]). Sodium and calcium channels have a comparable structure according to sequence homology [71]. The central pore of annexin V is structurally very different from the

**Fig. 7. Stereoview of the central channel formed by helices IIA, IIB, IVA and IVB seen from the side.**

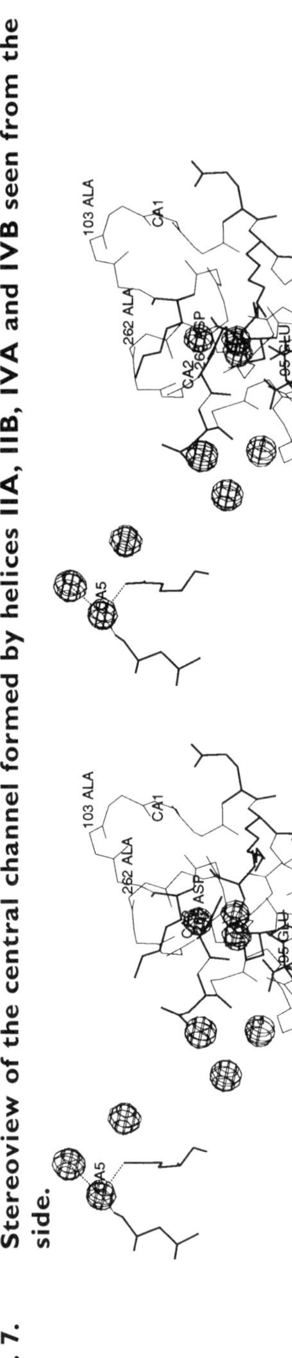


*The polar side chains projecting into it are drawn with thick lines. The chain of well localized bound water molecules found in the crystals is indicated by spheres.*

putative potassium channel pore. However, similarly highly charged peptides derived from the nicotinic acetylcholine receptor sequence have been found to form discrete conductance ion channels [72].

A change in the lifetime of open and closed states is the basis of transmembrane voltage-dependent gating and implies the presence of a voltage sensor, i.e. charges or equivalent dipoles, which move in an electric field (see [73], for a review). In annexin V the open state is favoured with increasing negative (and less so positive) intracellular holding potentials [56a]. Our crystal structure shows that annexin V is strongly electrically dipolar with the positive and negative poles located inside the modules (II, III) and (I, IV), respectively (Fig. 1 in [31]). An electric field may cause module motion as seen, when the $P6_3$ and R3 crystal structures [30,31] are compared. The different relative orientations could change the diameter of the central cavity and thereby switch the annexin V ion channel from a closed to an open state. Similar tilting motions have been envisioned for the penta-helical pore of the nicotinic acetylcholine receptor channel [74].

An external electric field may also induce rearrangement of the charged amino acid residues surrounding the pore in annexin V, which is an incision between modules (II, III) and (I, IV). The existence of two life times, mentioned above, implies two conformational transitions leading to opening/closing of the channel. We proposed a mechanical model in agreement with structural and kinetic data, in which the existence of one open and at least two different closed states is supposed. The open–closed transitions are regulated by independent opening/closing of two gates [56a].

The funnel-shaped pore of annexin V is formed by four α-helices and is wide open at the upper end, where it is in contact with the membrane and constricted at the opposite end. A funnel-shaped bundle of four amphipathic α-helices was recently conjectured as a plausible structural motif for a calcium-channel pore [75]. In annexin V this part of the molecule is formed by a large cavity confined between two ionic clusters, which we associate with the two gates from the mechanical model. Gate 1 is formed by the salt-bridge Arg-271–Glu-112 across the protein pore and the nearby Glu-95. The second gate (gate 2) is constituted by Arg-117 (forming a salt bridge with Asp-92) and Arg-276 (engaged in strong ionic interactions with Glu-121 and Asp-280). We demonstrated that there are possible conformations of the open state of the annexin V channel, in which the protein cavity is opened without drastic conformational changes of the side chains of the above-mentioned residues. The main effect of such rearrangements would be the exchange of the salt-bridge partners of two groups: Arg-271–Glu-112 to Arg-271–Glu-95 and Glu-121–Arg-276 to Glu-121–Arg-117. In the open state the charges of the participating residues remain in a similar and favourable electrostatic environment, which implies that the opening/closing of the gates is not accompanied by large changes of the electrostatic free energy. This conclusion is in accordance with the values we calculated from the open and closed life times for annexin V. The reciprocal values of the life times are the rate constants of the open/closed transitions [76]. These can be used to calculate the corresponding equilibrium constants and the Gibbs free energies. The resulting Gibbs free energies are, of course, voltage-dependent. With the increase of the holding potential in both directions, the open state is increasingly favoured but the Gibbs free energies do not exceed 1.5 kcal/mol [56a].

The conformational changes mentioned above may be coupled with the observed collective 'hinged domain motion' of the protein [31]. The calcium-rich structure (R3) shows a widening of the central protein pore, which suggests an enhancement of the flexibility of the charged groups in this region. Both transitions are energetically equivalent at negative voltages, but different at positive voltages. This asymmetry results from the asymmetry of the protein and its charge distribution.

Voltage-gated potassium channels occur in three states—closed, open and inactivated. The inactivated state is thought to be generated by a physical blockade of the ionic pore by the *N*-terminus of the protein, which was shown to be directly involved in this process ([77] and see [69] for a review). The close association of the *N*-terminus of annexin V with the central pore suggests that it plays a similar role. This is also evident in annexin II, where the *N*-terminus is the binding site for the small p11 protein [37,38].

Annexin VI (p68), the only known member of the annexin family with eight instead of four tandem repeats, modulates the $Ca^{2+}$-release-channel activity from the sarcoplasmic reticulum reconstituted in planar phospholipid bilayer membranes. Annexin VI itself shows no channel activity in the bilayer membranes [78]. We also could not establish annexin VI channel activity in planar phospholipid bilayer membranes (R. Berendes, unpublished work).

## References


1. Geisow, M. J. & Walker, J. H. (1986) Trends Biochem. Sci. **11**, 420–423
2. Crompton, M. R., Moss, S. E. & Crumpton, M. J. (1988) Cell **55**, 1–3
3. Klee, C. B. (1988) Biochemistry **27**, 6645–6653
4. Smith, V. L., Kaetzel, M. A. & Dedman, J. R. (1990) Cell Regul. **1**, 165–172
5. Geisow, M. J. (1991) Trends Biotechnol. **9**, 180–181
6. Moss, S. E., Edwards, H. C. & Crumpton, M. J. (1991) in Novel Calcium-binding Proteins (Heizmann, C. W., ed.), pp. 535–566, Springer-Verlag, Berlin
7. Kretsinger, R. H. (1980) CRC Crit. Rev. Biochem. **8**, 119–174
8. Crumpton, M. J. & Dedman, J. R. (1990) Nature (London) **345**, 212
9. Geisow, M. J. (1986) FEBS Lett. **203**, 99–103

10. Radke, K., Gilmore, T. & Martin, G. S. (1980) Cell **21**, 821–828
11. Erikson, E. & Erikson, R. L. (1980) Cell **21**, 829–836
12. Horseman, N. D. (1989) Mol. Endocrinol. **3**, 773–779
13. Gerke, V. (1989) FEBS Lett. **258**, 259–262
14. Walker, J. H. (1982) J. Neurochem. **39**, 815–823
15. Johnston, P. A., Perin, M. S., Reynolds, G. A., Wasserman, S. A. & Südhof, T. C. (1990) J. Biol. Chem. **265**, 11382–11388
16. Gerke, V. (1991) J. Biol. Chem. **266**, 1697–1700
17. Boustead, C. M., Smallwood, M., Small, H., Bowles, D. J. & Walker, J. H. (1989) FEBS Lett. **244**, 456–460
18. Smallwood, M., Keen, J. N. & Bowles, D. J. (1990) Biochem. J. **270**, 157–161
19. Geisow, M. J., Childs, J., Dash, B., Harris, A., Panayotou, G., Südhof, T. & Walker, J. (1984) EMBO J. **3**, 2969–2974
20. Südhof, T. C., Ebbecke, M., Walker, J. H., Fritsche, U. & Boustead, C. (1984) Biochemistry **23**, 1103–1109
21. Barton, G. J., Newman, R. H., Freemont, P. S. & Crumpton, M. J. (1991) Eur. J. Biochem. **198**, 749–760
22. Magendzo, K., Shirvan, A., Cultraro, C., Srivastava, M., Pollard, H. B. & Burns, A. L. (1991) J. Biol. Chem. **266**, 3228–3232
23. Pollard, H. B. & Rojas, E. (1988) Proc. Natl. Acad. Sci. U.S.A. **85**, 2974–2978
24. Burns, A. L., Magendzo, K., Shirvan, A., Srivasta, M., Rojas, E., Aligani, M. R. & Pollard, H. B. (1989) Proc. Natl. Acad. Sci. U.S.A. **86**, 3798–3802
25. Rojas, E., Pollard, H. B., Haigler, H. T., Parra, C. & Burns, A. L. (1990) J. Biol. Chem. **265**, 21207–21215
26. Newman, R., Tucker, A., Ferguson, C., Tsernoglou, D., Leonhard, K. & Crumpton, M. J. (1989) J. Mol. Biol. **206**, 213–219
27. Boustead, C. M., Walker, J. H., Kennedy, D. & Waller, D. A. (1991) FEBS Lett. **279**, 187–189
28. Lewit-Bentley, A., Doublie, S., Fourme, R. & Bodo, G. (1989) J. Mol. Biol. **210**, 875–876
29. Seaton, B. A., Head, J. F., Kaetzel, M. A. & Dedman, J. R. (1990) J. Biol. Chem. **265**, 4567–4569
30. Huber, R., Römisch, J. & Paques, E. (1990). EMBO J. **9**, 3867–3874
31. Huber, R., Schneider, M., Mayr, I., Römisch, J. & Paques, E. (1990) FEBS Lett. **275**, 15–21
32. Pollard, H. B., Burns, A. L. & Rojas, E. (1990) J. Membr. Biol. **117**, 101–112
33. Mosser, G., Ravanat, C., Freyssinet, J.-M. & Brisson, A. (1991) J. Mol. Biol. **217**, 241–245
34. Uzgiris, E. E. & Kornberg, R. D. (1983) Nature (London) **301**, 125–129
35. Brisson, A., Mosser, G. & Huber, R. (1991) J. Mol. Biol. **220**, 199–203
36. Döring, V., Schleicher, M. & Noegel, A. A. (1991) J. Biol. Chem., in the press
37. Johnsson, N., Marriott, G. & Weber, K. (1988) EMBO J. **7**, 2435–2442
38. Becker, T., Weber, K. & Johnsson, N. (1990) EMBO J. **9**, 4207–4213
39. Schlaepfer, D. D. & Haigler, H. T. (1987) J. Biol. Chem. **262**, 6931–6937
40. Ando, Y., Imamura, S., Hong, Y. M., Owada, M. K., Kakunaga, T. & Kannagi, R. (1989) J. Biol. Chem. **264**, 6948–6955
41. Powell, M. A. & Glenney, J. R. (1987) Biochem. J. **247**, 321–328
41a. Strynadka, N. C. J. & James, M. N. G. (1989) Annu. Rev. Biochem. **58**, 951–998
41b. Huber, R., Berendes, R. Burger, A. Schneider, M. Karshikov, A., Luecke, H., Roemisch, J. & Paques, E (1992) J. Mol. Biol. **223**, 683–704
42. Verheij, H. M., Volweik, J. J., Jansen, E. H. J. M., Puyk, W. C., Dijkstra, B. W., Drenth, J. & de Haas, G. H. (1980) Biochemistry **19**, 743–750
43. Thunnissen, M. M. G., Eiso, A. B., Kalk, K. H., Drenth, J., Dijkstra, B. W., Kuipers, O. P., Dijkman, R., de Haas, G. H. & Verheij, H. M. (1990) Nature (London) **347**, 689–691
44. Scott, D. L., White, S. P., Otwinowski, Z., Yuan, W., Gelb, M. H. & Sigler, P. B. (1990) Science **250**, 1541–1546

45. Gerke, V. & Weber, K. (1984) EMBO J. **3**, 227–233
46. Glenney, J. R. & Tack, B. F. (1985) Proc. Natl. Acad. Sci. U.S.A. **82**, 7884–7888
47. Johnsson, N., Johnsson, K. & Weber, K. (1988) FEBS Lett. **236**, 201–204
48. Johnsson, N., Vandekerckhove, J., Van Damme, J. & Weber, K. (1986) FEBS Lett. **198**, 361–364
49. Drust, D. S. & Creutz, C. E. (1988) Nature (London) **331**, 88–91
50. Meers, P. (1990) Biochemistry **29**, 3325–3330
51. Reutelingsperger, C. P. M., Kop, J. M. M., Hornstra, G. & Hemker, H. C. (1988) Eur. J. Biochem. **173**, 171–178
52. Tait, J. F., Gibson, D. & Fujikawa, K. (1989) J. Biol. Chem. **264**, 7944–7949
53. Möhwald, H. (1990) Annu. Rev. Phys. Chem. **41**, 441–476
54. Meers, P., Daleke, D., Hong, K. & Papahadjopoulos, D. (1991) Biochemistry **30**, 2903–2908
55. Andree, H. A. M., Reutelingsperger, C. P. M., Hauptmann, R., Hemker, H. C., Hermens, W. T. & Willems, G. M. (1990) J. Biol. Chem. **265**, 3923–3928
56. Schlaepfer, D. D., Mehlman, T., Burgess, W. H. & Haigler, H. T. (1987) Proc. Natl. Acad. Sci. U.S.A. **84**, 6078–6082
56a. Karshikov, A., Berendes, R., Burger, A., Cavalié, A., Lux, H. D. & Huber, R. (1992) Eur. Biophys. J. **20**, 337–344.
57. Sugar, I. P. & Neumann, E. (1984) Biophys. Chem. **19**, 211–225
58. Sugar, I. P., Förster, W. & Neumann, E. (1987) Biophys. Chem. **26**, 321–335
59. Neumann, E. (1988) Ferroelectrics **86**, 325–333
60. Deisenhofer, J., Epp, O., Miki, K., Huber, R. & Michel, H. (1985) Nature (London) **318**, 618–624
61. Milks, L. C., Kumar, N. M., Houghton, R., Unwin, N. & Gilula, N. B. (1988) EMBO J. **7**, 2967–2975
62. Changeux, J. P. (1990) Trends Pharmacol. Sci. **11**, 485–492
63. Stroud, R. M., McCarthy, M. P. & Shuster, M. (1990) Biochemistry **29**, 11009–11023
64. Henderson, R., Baldwin, J. M., Ceska, T. A., Zemlin, F., Beckmann, E. & Downing, K. H. (1990) J. Mol. Biol. **213**, 899–929
65. Reference deleted
66. Yool, A. J. & Schwarz, T. T. (1991) Nature (London) **349**, 700–704
67. Hartmann, H. A., Kirsch, G. E., Drewe, J. A., Taglialatela, M., Joho, R. H. & Brown, A. M. (1991) Science **251**, 942–944
68. Yellen, G., Jurman, M., Abramson, T. & MacKinnon, R. (1991) Science **251**, 939–942
69. Miller, C. (1991) Science **252**, 1092–1096
70. Stevens, C. F. (1991) Nature (London) **349**, 657–658
71. Numa, S. (1989) Harvey Lect. **83**, 121–165
72. Gosh, P. & Stroud, R. M. (1991) Biochemistry **30**, 3551–3557
73. Hille, B. (1984) Ionic Channels of Excitable Membranes. Sinauer Associates, Sunderland, MA
74. Galzi, J.-L., Revah, F., Basis, A. & Changeux, J. P. (1991) Annu. Rev. Pharmacol. **31**, 37–72
75. Grove, A., Tomich, J. M. & Montal, M. (1991) Proc. Natl. Acad. Sci. U.S.A. **88**, 6418–6422
76. Colquhoun, D. & Hawkes, A. G. (1981) Proc. R. Soc. London B. **112**, 205–235
77. Zagotta, W. N., Hoshi, T. & Aldrich, R. W. (1990) Science **250**, 568–571
78. Diaz-Munoz, M., Hamilton, S. L., Kaetzel, M. A., Hazarika, P. & Dedman, J. R. (1990) J. Biol. Chem. **265**, 15894–15899

11

# Annexin VI

**John R. Dedman and Marcia A. Kaetzel**

Department of Physiology and Biophysics, University of Cincinnati College of Medicine, 231 Bethesda Avenue, Cincinnati, OH 45267-0576, U.S.A.

## Introduction

The recognition of calcium as an important physiological regulator began with the studies of Ringer [1] in which he described the requirement for calcium in the maintenance of contraction of isolated frog hearts. Hodgkin and Keynes [2] postulated a mobile intracellular calcium receptor when they found that $^{45}Ca^{2+}$ microinjected into giant squid axons did not act as a free ion in an electrical field. Further physiological studies established the importance of calcium in the release of secretory granules [3–5] and in the regulation of glycogenolysis [6]. These studies promoted the concept that calcium had specific targets to carry out its intracellular function. In an effort to identify soluble intracellular calcium receptors, Wolff and Siegel [7] were the first to purify calmodulin. The initial demonstration of a role for calcium in association with specific target proteins culminated in the studies of Ebashi and Endo [8] and Greaser and Gergely [9]. They identified troponin as a calcium-responsive regulator of muscle contraction.

In retrospect, a fascinating aspect of the annexins is that, in spite of their abundance and wide-spread distribution and the intense interest in finding intracellular mediators, they were not identified until much later than calmodulin. Because of their calcium-dependent phospholipid binding properties, annexins have been identified repeatedly by independent approaches. Creutz *et al.* [10] initially described a protein that potentiated chromaffin granule fusion, and termed it synexin. In 1981, Creutz *et al.* [11] reported further on the large number of polypeptides that bind to chromaffin granule membranes in a $Ca^{2+}$-dependent manner. Likewise, Geisow and Burgoyne [12] studied the calcium- and calmodulin-dependent association of proteins with isolated chromaffin granule membranes; the annexins were among this complex mixture of proteins. Independently, Owens *et al.*, [13] and Walker [14] identified members of the annexin family through $Ca^{2+}$-dependent binding to detergent-extracted membranes. Moore and Dedman [15], using a different approach, found that proteins in addition to calmodulin bound to fluphenazine–Sepharose resins in a $Ca^{2+}$-dependent manner. In addition, Weber and colleagues [16] identified proteins associated with intestinal microvilli. As a result of these independent discoveries premature assignment of function occurred. Also contributory was the fact that many of these proteins and their proteolytic products had similar relative molecular masses and antibody cross-reactivity. Confusion in terminology resulted. Molecular cloning and cDNA sequencing of each member revealed a related protein family and established a basis for precise assignment of chemical identity. This sequence information allowed for the implementation of a single terminology, thus improving communication between laboratory groups [17]. Through all of these co-discoveries and controversies over exact identity of the four-domain annexins, there was little argument between laboratories regarding annexin VI, as it migrated as a 67–73 kDa protein on SDS/PAGE. The parental molecule, however, is very susceptible to proteases [18] and the 37 kDa 'core' was probably detected and confused as an independent protein in the earlier reports on the annexins. Monospecific antibodies have proved valuable in resolving this problem [19].

## Structure and properties

Human and mouse annexin VI cDNAs have been cloned and sequenced [20–22]. The derived amino acid sequence predicts a polypeptide of 673 residues with a 95% similarity between the two species. As has been shown for the other annexins, annexin VI is composed of repeated domains of 65–70 amino acids, indicating gene duplications. Annexin VI, in contrast to all of the other annexins, contains an additional duplication producing an eight-domain structure. The other distinctive feature when comparing the annexins is the *N*-terminal tail which is hypervariable in sequence and length. This region is readily accessible to protein kinases and proteases. Annexin VI contains a comparatively short *N*-terminal tail and respective linker sequence between the two four-domain cores. Many of the amino acids within the domains are highly conserved between all of the annexins suggesting sites of similar function such as calcium- and phospholipid-binding. We have shown that a site-directed antibody against KAMKGLGTDE recognizes annexins I–VI [23,24]. Analysis of human annexin V crystals at the 2 Å level indicates that the amino acids that form the calcium-binding pocket are discontinuous. The proposed calcium-coordinating structure contains the sequence MKG(A or L)GT(39 residues)(D or E) [25]. This prototype sequence is found in all annexins in domains 2 and 4 but has not been found in the reported domain 3 sequences. Annexin VI contains six potential calcium-binding sites in domains 1, 2, 4, 5, 6 and 8 [25].

Crompton *et al.* [21] reported finding two distinct cDNAs for annexin VI. One lacks the sequence VAAEIL in domain seven; this deletion also occurs in mouse [22]. Potentially an alternative splicing event may generate these two cDNAs and may account for the 67 kDa doublet seen on SDS/PAGE for annexin VI. The physiological significance of the annexin VI isoforms is not known.

Annexin VI exists as a monomer in the absence of phospholipids and binds 1–4 moles of calcium with micromolar affinity [13,26,27,28]. In the presence of phospholipid, however, annexin VI has been reported to bind 8 moles of calcium per mole of protein [28]. The combination of calcium and phospholipid causes self-association of the protein to form a complex [29,30]. This type of protein–protein interaction has been highly resolved for annexin V by Brisson *et al.* [31] using low-resolution two-dimensional X-ray crystallography. Annexin V forms triskelions which, in turn, self-associate into a regular lattice which covers the phospholipid bilayer surface. *In vivo*, this $Ca^{2+}$-dependent macromolecular arrangement would alter the membrane properties by sequestering specific phospholipids and would alter the membrane fluidity. This interaction would then modify the activities of proteins such as pumps and channels, which are associated with the membranes. An additional function for phospholipid binding may be the targeting of individual annexins to specific cellular membranes. Edwards and Crumpton [32] used lipid immobilized onto phenyl-Sepharose to demonstrate a hierarchy in $Ca^{2+}$-dependent phospholipid binding for annexin VI. The free calcium concentrations required for half-maximal binding were 0.4 mM for phosphatidylethanolamine, 0.6 mM for phosphatidylserine and 3.5 mM for phosphatidylinositol. These values are an order of magnitude less than those found for annexin IV under the same conditions. Annexin VI was also shown to bind arachidonic, palmitic and oleic acids [32]. Reconstitution of highly purified components as performed in these *in vitro* studies may only hint at the actual annexin properties in the dynamic, complex situation under physiological conditions. For example, Drust and Creutz [33] have shown that the $Ca^{2+}$ required to fuse vesicles by annexin II is markedly reduced when comparing unfractionated membranes with chemically defined liposomes. Understanding the mechanism of action of annexin VI will require identification of all associated ligands involved in cellular targeting and regulation. The cellular regulation of protein kinase C is a useful paradigm for understanding the annexins. The presence of phosphatidylserine and diacylglycerol markedly reduces the $Ca^{2+}$ requirement for enzyme activity [34]. In addition, the cell contains proteins that potentially target the kinase to specific cellular sites. These receptors for activated protein kinase C alter the protein substrate specificity [35]. In a similar manner, binding to specific lipids and targeting proteins and post-translational modifications such as phosphorylation may all contribute to the physiological regulatory role of annexin VI.

## Cellular function

Most of the assignments of function to the annexins have evolved for their physiochemical properties such as their ability to bind specific phospholipids, membranes or cytoskeleton, and to act as substrates for specific protein kinases. Our approach to elucidating the cellular function of the annexins has been to localize each protein in a tissue with highly specialized functions and to evaluate activities in biochemically reconstituted systems. We first produced monoclonal and affinity-purified polyclonal antibodies [19]. These antibodies were then used to evaluate expression and subcellular localization. This approach allows correlation of annexin expression with known calcium-dependent subcellular functions. Efforts can then be focused on defined biochemical reconstitution studies. Immunoblots of total tissue extracts indicated that annexin VI was differentially expressed from tissue to tissue and that muscle contained high concentrations of the protein (Fig. 1) [19,36]. This finding warranted localization studies in skeletal muscle. Immunolocalization of annexin VI in cultured myotubes demonstrated a regular striated pattern of expression (Fig. 2). Antibody staining of muscle fibre cross-sections was around individual myofibrils ([37]; see also Fig. 3). This reticular pattern provided strong evidence that annexin VI was not localized to the sarcolemma or the contractile elements, but was more suggestive of an association with the sarcoplasmic reticulum. This highly differentiated organelle releases $Ca^{2+}$ following membrane depolarization to induce contraction, and then resequesters $Ca^{2+}$ to cause relaxation. Studies using intact, isolated vesicles from skeletal muscle sarcoplasmic reticulum did not reproducibly demonstrate an alteration in biochemical function, $Ca^{2+}$ uptake or $Ca^{2+}$ release, following the addition of annexin VI or anti-annexin VI [37]. Annexin VI was then tested in electrophysiological studies. Vesicles enriched in the ryanodine-sensitive $Ca^{2+}$-release channel isolated from rabbit skeletal muscle sarcoplasmic reticulum were inserted into artificial lipid bilayers. This technique allows for the measurement of several conductance properties of individual channels incorporated into the bilayer. Annexin VI, when added at concentrations of between 6 and 60 nM

**Fig. 1. Immunoblot analysis of total tissue extracts with monospecific annexin VI antibody.**

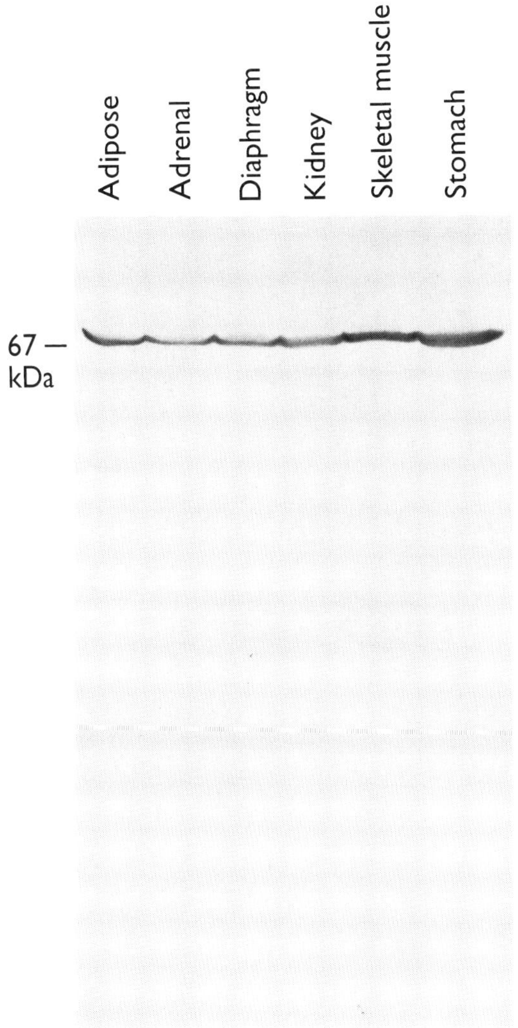


*Tissues were solubilized in EDTA, 2-mercaptoethanol and SDS according to Kaetzel, Hazarika & Dedman [24]. Total protein (100 µg) was separated by SDS-PAGE and electroblotted onto nitrocellulose. The sheet was incubated with affinity-purified sheep annexin VI antibody followed by phosphate-buffered saline washings and peroxidase-conjugated rabbit anti-sheep IgG secondary antibody. Detection of this complex was accomplished with 4-chloro-1-naphthol.*

produced striking changes in the gating properties of the calcium-release channel (Fig. 4a and d). The open probability ($P_0$) of the channel increases from approximately 30 % to 75 % [38]. The mean open time constant was more striking. The untreated channel in the bilayer produces a chaotic

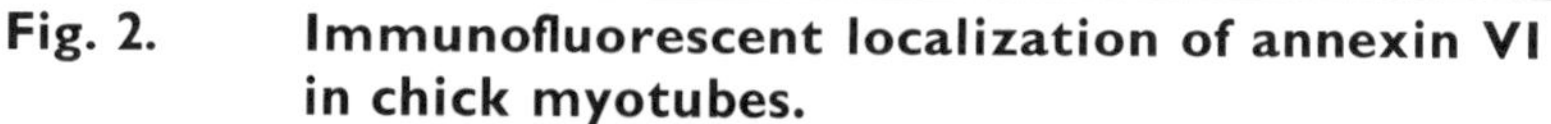

**Fig. 2. Immunofluorescent localization of annexin VI in chick myotubes.**

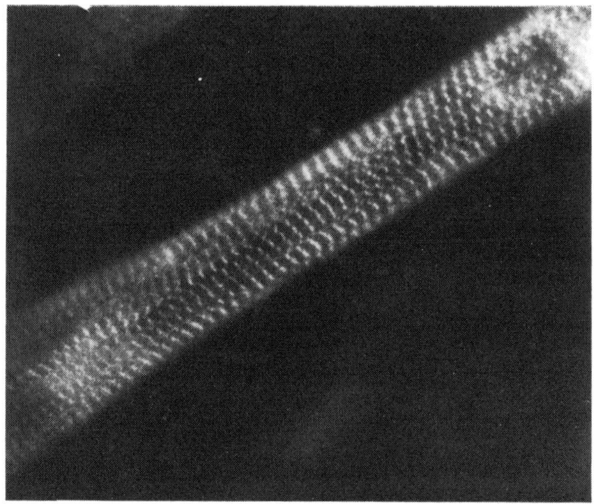

*Well developed myotubes cultured on glass coverslips were fixed in methanol and incubated successively with sheep anti-annexin VI antibody, phosphate-buffered saline and fluorescein-conjugated rabbit anti-sheep IgG. The images were recorded on Kodak T-MAX (ASA 100) using a Nikon Optiphot epifluorescence microscope.*

'flicker' of rapid openings and closures of 0.6 ms. Addition of annexin VI increased the mean open time by greater than 50-fold. In order to determine whether the annexin VI effect on the $Ca^{2+}$-release channel was $Ca^{2+}$- dependent, the free $Ca^{2+}$ concentration was reduced to 0.05 μM with the addition of EGTA. The channel returned to the pre-annexin VI-treated state, gating rapidly with very brief open and close times. Channel properties returned to long mean open times by the readdition of 20 μM free calcium. Preliminary studies indicate that annexins I–V do not alter the gating behaviour of the release channel. Furthermore, the *N*-terminal or *C*-terminal halves do not modify the channel activity. The parental protein is bisected with V8 protease at position 334 which is in the linkage region between domains 4 and 5. This cleavage produces two different four-domain halves of almost equal molecular weight ([18]; see also Fig.

5). Each protease-resistant core was purified by Mono Q fast-protein liquid chromatography and shown to retain the ability to bind phospholipid in a $Ca^{2+}$-dependent manner. Neither half of the annexin VI molecule, however, alone or in combination, affected the release channel ([18]; see

**Fig. 3. Immunofluorescent localization of annexin VI in skeletal muscle cross-sections.**

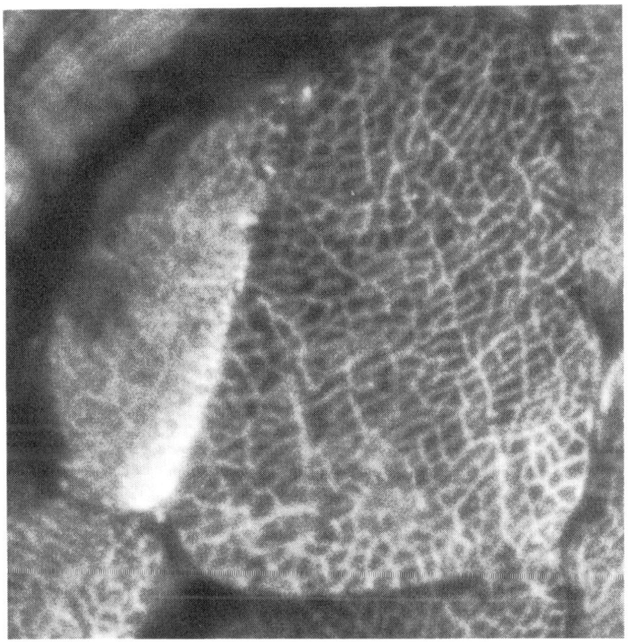

*Rat oesophagous muscle was fixed in 10% formalin, embedded in paraffin and sectioned at 4 µm. The deparaffinized sections were incubated with a monospecific monoclonal anti-annexin VI antibody, washed with phosphate-buffered saline and incubated with fluorescein-conjugated goat anti-mouse IgG. The field was photographed on Kodak T-MAX (ASA 100) film.*

also Fig. 4b and c). An additional aspect of this annexin VI regulation is sidedness of the response. Annexin VI appears to be effective only from the ATP-insensitive side of the channel in the bilayer. Differential extraction of isolated intact heavy sarcoplasmic reticulum vesicles with EGTA and 1% Triton X-100 demonstrated that the intact vesicles contain annexin VI in the lumen [38].

Immunolocalization of annexin VI in neutrophils indicated staining of punctate structures throughout the cytoplasm (Fig. 6).

**Fig. 4. Modulation of the sarcoplasmic reticulum $Ca^{2+}$-release channel by intact annexin VI.**

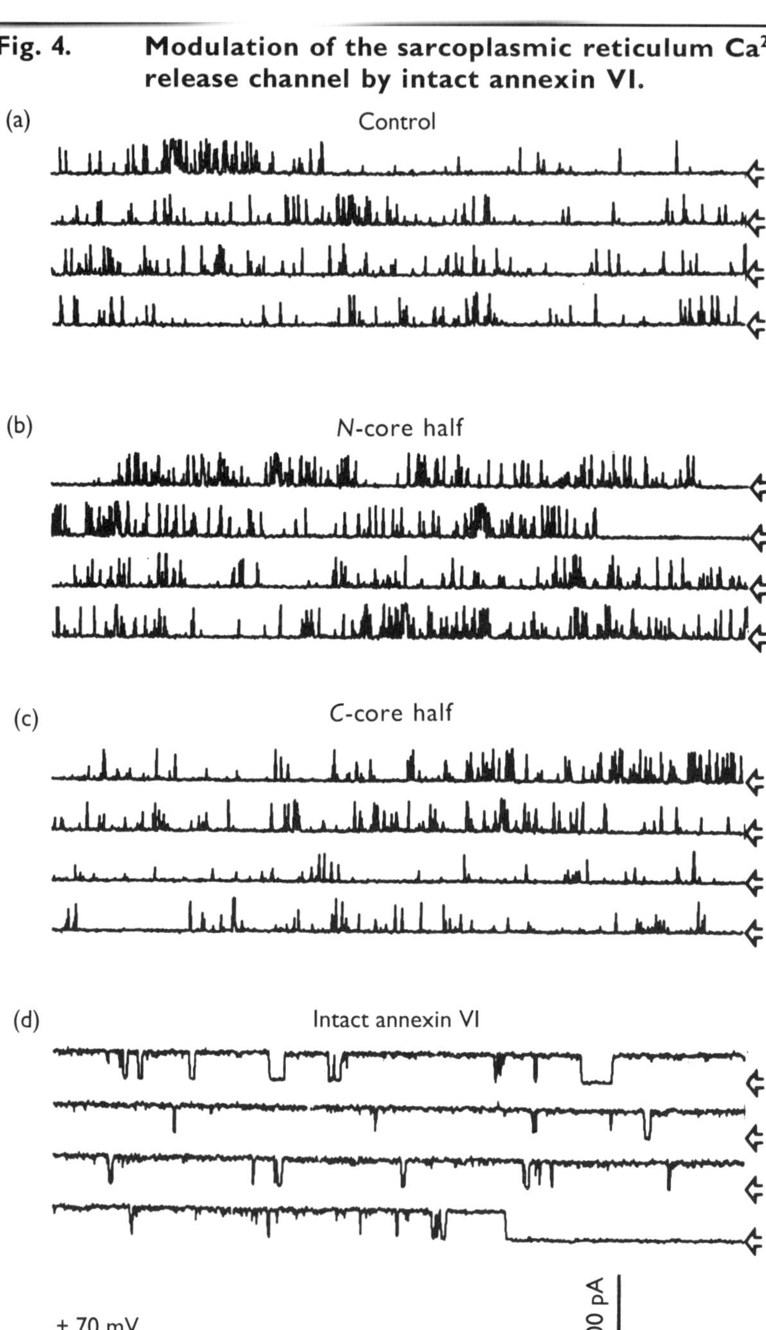

**Fig. 5. Proteolytic digestion of annexin VI to produce resistant 37 kDa cores.**

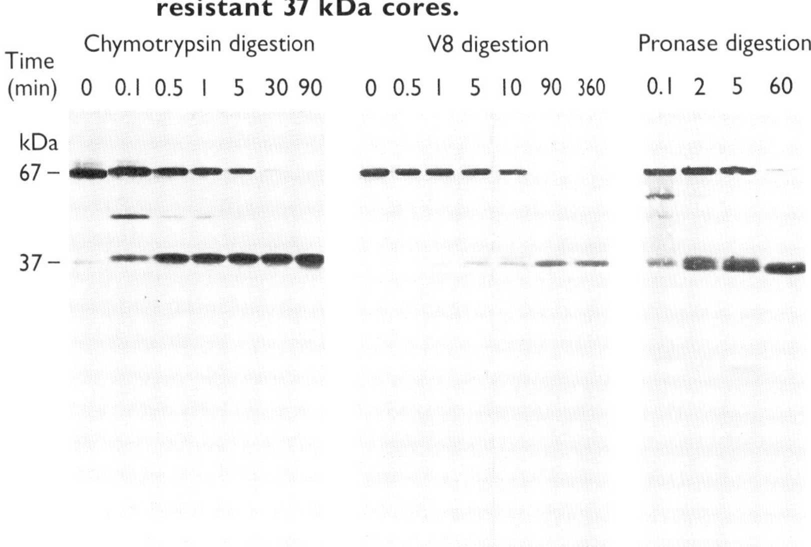


*Annexin VI (20 μg) was treated with 0.1 μg chymotrypsin, 4 μg* Staphylococcus *V8 protease or 0.01 μg Pronase E. At the times indicated an aliquot was removed and the reaction terminated by heating at 90 °C for 10 min in 10 mM EDTA, 1% 2-mercaptoethanol and 5% SDS.*

Likewise, annexin VI antibody staining of an undifferentiated muscle cell-line expressing chick muscle $Ca^{2+}$-ATPase demonstrated well defined hollow spherical structures ranging from less than 0.1 μm to 0.5 μm in diameter (Fig. 7). At the light microscope level, it could not be determined whether annexin VI was lining the inner or outer surface or both surfaces of these intracellular vesicles. These results raise the possibility that annexin VI is associated with membranous intracellular organelles such as the

*The records were obtained from a planar lipid bilayer system described by Diaz-Muñoz et al. [43]. Microsomal membranes obtained from skeletal muscle were fused with the bilayer by gentle stirring. The arrows indicate the closed channel baseline. Preparations are as follows: (a) control; (b) 60 nM purified N-terminal 37 kDa core; (c) further addition of 60 nM purified C-terminal 37 kDa core to the same preparation; and (d) further addition of 6 nM intact annexin VI to the same preparation.*

endoplasmic reticulum. A primary question is what cellular mechanism is employed to transport the protein across the sarcoplasmic reticulum or endoplasmic reticulum membrane, as the annexins do not have traditional *N*-terminal signal sequences. Immuno-electron microscopy localization of

**Fig. 6. Immunofluorescent localization of annexin VI in Raab neutrophils.**

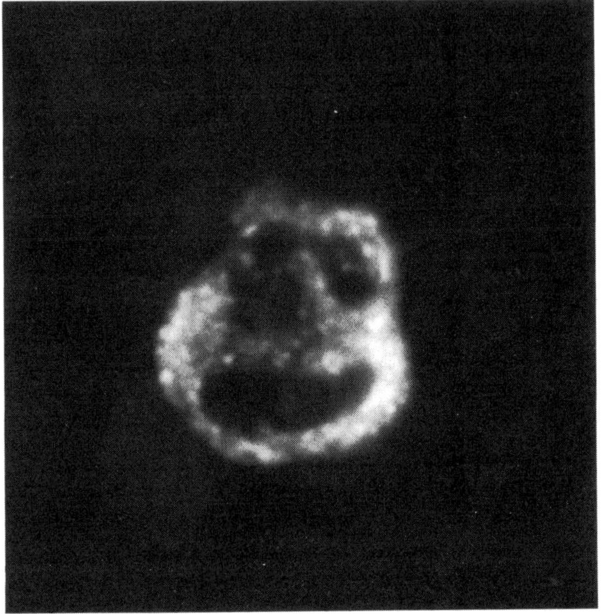

*Air-dried human blood smears were lysed in −20 °C acetone for 10 min and then processed as described in Fig. 2.*

annexin VI using monospecific affinity-purified antibodies should indicate whether some fraction of the expressed protein is within the sarcoplasmic reticulum lumen. Although the annexins do not possess *N*-terminal signal sequences, movement into and across membranes may be associated with the ability of the annexins to bind and fuse membrane vesicles.

Most of the studies reported to date have attempted to associate biochemical properties with tissue expression of the annexins in order to postulate cellular functions. Ideally, in order to confirm such hypotheses, annexin function should be evaluated in living cells. Such an endeavour is not a trivial undertaking, especially without high-affinity, specific inhibitors or null mutants. A great advantage would be the ability to control the genetic make-up and expression of specific proteins, as is possible in yeast and fungi. A question currently unanswered is whether these lower forms of life can be used to correlate the functions of annexins in differentiated mammalian cells. It is encouraging that an analogue of annexin VII is expressed in the slime mould *Dictyostelium discoideum* [39].

**Fig. 7. Immunofluorescent localization of annexin VI in a mouse muscle cell line expressing chick muscle $Ca^{2+}$-ATPase.**

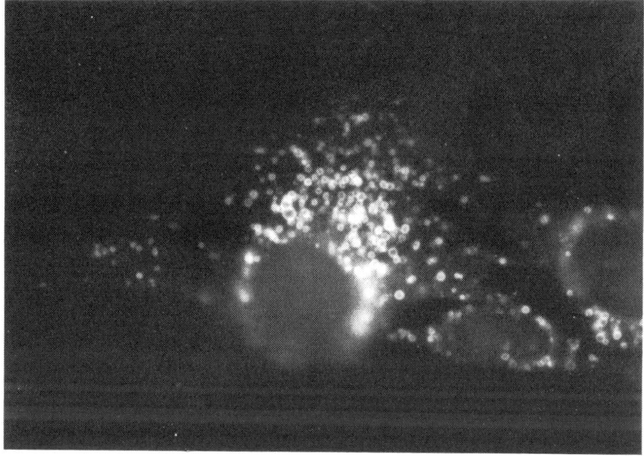

*The cell line, C2FCa2, was a gift of N. Karin (University Texas Medical School, Houston). This transfected cell line expresses the chicken skeletal muscle $Ca^{2+}$-ATPase [43]. The cells were processed as described in Fig. 2.*

This cellular system has proven useful for studies involving recombination of homologous genes. Therefore, once the annexin VI gene is cloned and mutated it can be used to disrupt and replace the endogenous gene by homologous recombination and the resultant physiological changes can be evaluated [40]. More complex eukaryotic cellular systems that do not have a haploid stage offer the potential for genetic manipulation, but the frequency of homologous recombination is extremely low [41]. In order to elucidate the physiological role of annexin VI in complex eukaryotes the gene must be isolated. Selectable homologous recombination can then be attempted in pluripotent embryonic stem cell lines established from normal mouse embryos which will resume normal embryonic development following implantation in a foster mother. The role of specific annexin VI mutations could be evaluated in a living organism.

Until such technology is routine, alternative approaches must be attempted. For example, the cellular function of annexin VI can be evaluated following selective inhibition of mRNA translation. This suppression can be accomplished by oligonucleotides that are complementary to a target cellular mRNA [42,43]. It has been demonstrated that synthetic phosphothioate oligonucleotides are relatively permeable to cellular membranes, stable to nucleases and able to inhibit specific mRNA translation. We have used phosphothioate oligonucleotides anti-sense to the reported mouse mRNA sequence to selectively inhibit the synthesis and in turn reduce the cellular levels of annexin VI. We have found that morphology and survival of differentiated cultured muscle cells are not altered when the annexin VI is reduced to 25% of levels present in untreated cells. In light of our studies with the sarcoplasmic reticulum calcium-release channel we intend to evaluate changes in contractile properties and fura-II fluorescence following stimulation of differentiated muscle cells. These studies may provide further evidence relating annexin VI to the regulation of intracellular calcium homeostasis.

## References


1. Ringer, S. (1882) J. Physiol. (London) **4**, 29
2. Hodgkin, A. L. & Keynes, R. D. (1957) J. Physiol. (London) **138**, 253–281
3. Douglas, W. W. & Rubin, R. P. (1961) J. Physiol. (London) **159**, 40–57
4. Eccles, J. C. (1964) The Physiology of Synaptases, Springer-Verlag, Berlin
5. Katz, B. & Miledi, R. (1968) J. Physiol. (London) **195**, 481–492
6. Meyer, W. L., Fischer, W. H. & Krebs, E. G. (1964) Biochemistry **3**, 1033–1039
7. Wolff, D. J. & Siegel, F. L. (1972) J. Biol. Chem. **247**, 4180–4185
8. Ebashi, S. & Endo, M. (1968) Prog. Biophys. Mol. Biol. **18**, 123
9. Greaser, M. L. & Gergely, J. (1971) J. Biol. Chem. **2466**, 4226–4233
10. Creutz, C. E., Pazoles, C. J. & Pollard, H. B. (1978) J. Biol. Chem. **253**, 2858–2866
11. Creutz, C. E. (1981) Biochem. Biophys. Res. Commun. **103**, 1395–1400
12. Geisow, M. J. & Burgoyne, R. D. (1982) J. Neurochem. **38**, 1735
13. Owens, R. J., Gallagher, C. J. & Crumpton, M. J. (1984) EMBO J. **3**, 945
14. Walker, J. H. (1982) J. Neurochem. **39**, 815
15. Moore, P. & Dedman, J. (1982) J. Biol. Chem. **257**, 9663–9667
16. Shadle, P. J., Gerke, V. & Weber, K. (1985) J. Biol. Chem. **260**, 16354
17. Crumpton, M. J. & Dedman, J. R. (1990) Nature (London) **345**, 212
18. Hazarika, P., Sheldon, A., Kaetzel, M. A., Diaz-Muñoz, M., Hamilton, S. L. & Dedman, J. R. (1991) J. Cell. Biochem. **46**, 86–93
19. Kaetzel, M. A., Hazarika, P. & Dedman, J. R. (1900) in Stimulus Response Coupling: the Role of Intracellular Calcium-binding Proteins (Smith, V. L. & Dedman, J. R., eds.), pp. 383–409, CRC Press, Boca Raton
20. Sudhof, T. C., Slaughter, C. A., Leznicki, I., Barjon, P. & Reynolds, G. A. (1988) Proc. Natl. Acad. Sci. U.S.A. **85**, 664–668
21. Crompton, M. R., Owen, R. J., Totty, N. F., Moss, S. E., Waterfield, M. D. & Crumpton, M. J. (1988) EMBO J. **7**, 21–27
22. Moss, S. E., Crompton, M. R. & Crumpton, M. J. (1988) Eur. J. Biochem. **177**, 21–27
23. Kaetzel, M. A. & Dedman, J. R. (1989) Biochem. Biophys. Res. Commun. **160**, 1233–1237
24. Kaetzel, M. A., Hàzarika, P. & Dedman, J. R. (1989) J. Biol. Chem. **264**, 14463–14470
25. Huber, R., Schneider, M., Mayr, I., Romisch, J. & Paques, E.-P. (1990) FEBS Lett. **275**, 15–21
26. Matthew, J. K., Krolak, J. M. & Dedman, J. R. (1986) J. Cell Biochem. **32**, 223–234
27. Mani, R. S. & Kay, C. M. (1989) Biochem. J. **259**, 799

28. Yoshizaki, H., Arai, K., Mizoguchi, T., Shiratsuchi, M., Hattori, Y., Nagoya, T., Shidara, Y. & Maki, M. (1989) J. Biochem. (Tokyo) **105**, 178–183
29. Newman, R., Tucker, A., Ferguson, C., Tsernoglou, D., Leonard, K. & Crumpton, M. J. (1989) J. Mol. Biol. **206**, 213–219
30. Zaks, W. J. & Creutz, C. B. (1991) Biochemistry **30**, 9607–9615
31. Brisson, A., Mosser, G. & Huber, R. (1991) J. Mol. Biol. **220**, 199–203
32. Edwards, H. C. & Crumpton, M. J. (1991) Eur. J. Biochem. **198**, 121–129
33. Drust, D. S. & Creutz, C. E. (1988) Nature (London) **331**, 88–91
34. Nishizuka, Y. (1988) Nature (London) **334**, 661–665
35. Mochly-Rosen, D., Khaner, H., Lopez, J. & Smith, B. L. (1991) J. Biol. Chem. **266**, 14866–14868
36. Smith, V. L. & Dedman, J. R. (1986) J. Biol. Chem. **261**, 15815–15818
37. Hazarika, P., Kaetzel, M. A., Sheldon, A., Karin, N. J., Fleischer, S., Nelson, T. E. & Dedman, J. R. (1991) J. Cell. Biochem. **46**, 78–85
38. Diaz-Muñoz, M., Hamilton, S. L., Kaetzel, M. A., Hazarika, P. & Dedman, J. R. (1990) J. Biol. Chem. **265**, 15894–15899
39. Gerke, V. (1991) J. Biol. Chem. **266**, 1697–1700
40. DeLozanne, A. (1987) in Methods in Cell Biology (Spudich, J. A., ed.) **38**, 489–495, Academic Press, Orlando, FL
41. Frohman, M. A. & Martin, G. R. (1989) Cell **56**, 145–147
42. Sunikawa, K. & Miledi, R. (1988) Proc. Natl. Acad. Sci. U.S.A. **85**, 1302–1306
43. Marcus-Sekura, C. J. (1988) Anal. Biochem. **172**, 289–295
44. Karin, N., Kaprielian, Z. & Fambrough, D. (1989) Mol. Cell. Biol. **9**, 1978–1986

12

# Molecular biology and biochemistry of annexins V and VIII

**Rudolf Hauptmann* and Chris P. M. Reutelingsperger†**

*Ernst Boehringer Institut für Arzneimittelforschung, der Fa. Bender + Co, Dr. Boehringergasse 5-11, A-1121 Vienna, Austria
†Department of Biochemistry, University of Limburg, P.O. Box 616, NL-6200 MD Maastricht, The Netherlands

## Introduction

Annexin V was first discovered in our laboratory during the measurement of the procoagulant activity of a fractionated homogenate of human umbilical cord arteries. The soluble fraction, containing proteins with molecular masses ranging from 30–40 kDa, inhibited the prothrombin time. The anticoagulant activity was demonstrated to be associated with a protein of 32 kDa [1] and appeared to be based on a mechanism not yet described in the field of coagulation. The protein was named vascular anticoagulant (VAC) and was further characterized. No proteolytic and no serpin activity could be attributed to VAC and its inhibitory effect on procoagulant reactions could be explained by its $Ca^{2+}$ dependent high-affinity phospholipid-binding properties [1,2]. Through means of molecular biology the primary structure of VAC was elucidated [3].

Meanwhile, other laboratories discovered the same protein using coagulation assays, phospholipase $A_2$ assays or protein analysis, resulting in a diverse nomenclature. VAC was found to be identical to inhibitor of blood coagulation or calphobindin-I [4], endonexin II [5], placental protein 4 [6], placental anticoagulant protein I [7], lipocortin V [8], 35 $\gamma$-calcimedin [9] and anchorin CII [10]. The primary structure revealed VAC to be a member of a family of proteins sharing structural and functional features and named annexins [11,12]. Recently, a consensus nomenclature was introduced and VAC was termed annexin V [13].

During our initial screening of a human placenta cDNA library with a probe specific for annexin V, we isolated three overlapping cDNA clones that were derived from a mRNA encoding a novel annexin which was named VAC-β [14] and, according to the new nomenclature, annexin VIII [13].

This chapter describes in more detail the molecular biology of annexins V and VIII, as well as aspects of protein structure and *in vitro* activities, that should provide insights into their physiological importance, an area not yet clearly understood for the annexins despite the wealth of accumulated data.

## Human annexin V cDNA cloning

Annexin V cDNA was isolated either by oligonucleotide screening of placental and peripheral blood lymphocyte cDNA libraries [3,5,6] or by using monoclonal [4] or polyclonal [7] antibodies to screen placental expression cDNA libraries. The longest cDNA isolated comprises 1560 bp and contains 128 5′-non-translated nucleotides with an extremely high G/C content (70%). Kaplan *et al.* [5] described a 159 bp 5′-terminal end but the first 31 bases seemed to be derived from linker or vector sequences. Primer extension on poly(A)$^+$ RNA revealed a length of 153–157 nucleotides for the 5′-non-translated part of the mRNA [3]. The context of the first AUG is highly homologous to the Kozak consensus sequence [15] and is at the beginning of a 960 bp open reading frame encoding 320 amino acids. The reading frame is followed by 472 non-translated bases containing a canonical AATAAA polyadenylation signal, 22 bases upstream of the poly(A) tail. The entire length of the mRNA should therefore amount to 1589 bases plus the poly(A) tail. This is in good agreement with the estimated length of 1700–1800 bases derived from Northern blot experiments with RNA from various tissues and cell lines, where annexin V cDNA hybridizes to a single mRNA species [3,6,8].

The encoded protein comprises 319 amino acids after proteolytic removal of the start methionine with a calculated molecular mass of 35765 Da. The *N*-terminal alanine is blocked by acetylation [4,7]. Annexin V contains 30.7% charged amino acids, 54 acidic (aspartic and glutamic acid) and 41 basic (lysine and arginine) amino acids resulting in an isoelectric point of 4.8–4.9 [3,4]. Annexin V is the only member of the annexin family containing a single cysteine. This cysteine may be involved in some intermolecular disulphide bond formation, since about 2–4% of natural annexin V migrates as a dimer under non-reducing conditions on polyacrylamide gels. It does not contain an N-linked glycosylation site (Asn-Xaa-Ser/Thr). The formerly reported 2.4% carbohydrate content of annexin V [16] must therefore be attributed to O-glycosylation or may be derived from impurities of early annexin V preparations [6]. There is no signal sequence or other hydrophobic stretch long enough to span a membrane.

Annexin V cDNA is expressed in and the recombinant protein isolated from bacteria at high yields [3–5]. The recombinant protein is processed, meaning that the *N*-terminal formyl-methionyl residue is completely removed. Comparison of the natural and recombinant protein showed no difference in the physicochemical and biochemical properties of the two proteins except the isoelectric point being 0.1 pI units higher for the recombinant protein, owing to the lack of *N*-terminal acetylation in bacteria.

As a member of the $Ca^{2+}$/phospholipid-binding protein family, annexin V shows the typical fourfold repeat of 67/68 amino acids containing a highly conserved 17-amino-acid consensus sequence, recognized as a common feature probably important for $Ca^{2+}$/phospholipid-binding in annexin-like proteins [11,17], with spacers of variable length connecting the repeats. Comparison with the other annexins reveals annexin V as having the shortest annexin-specific *N*-terminus. In contrast to annexins I, II and IV, annexin V is not a substrate for protein kinase C [5] despite a threonine at position 7, possibly because of the lack of a lysine two amino acids downstream. Another explanation for not being phosphorylated by this enzyme is the truncated length of the *N*-terminal domain. Like annexins I and II [18], annexin V binds to actin in the presence of high $Ca^{2+}$ concentrations. Gly-74, -146, -230 and -305 near the ends of the individual repeats may be responsible for this binding. These glycines lie in the centre of a region sharing sequence similarity with gelsolin, another $Ca^{2+}$-dependent actin-binding protein [19].

## Animal annexin V cDNA cloning

Two animal annexin V cDNAs have been cloned from rat [8] and chicken (also named anchorin CII; [20] and correction in [10]). The cDNA-derived amino acid sequence of rat annexin V is 91.5 % identical to the human sequence, and the chicken sequence 79.0 % identical. Northern analysis with the 5′ half of the avian annexin V cDNA detects a single 1.7 kb mRNA whereas the 3′ part in addition hybridizes to a 5 kb mRNA species of unknown origin. Anchorin CII was identified as a collagen type II-binding protein on the surface of chondrocytes [21,22]. The mechanism by which anchorin CII appears on the surface of chick chondrocytes or skin fibroblasts remains unclear. Like human and rat annexin V, chicken anchorin CII lacks any sequence long enough to span a membrane.

## Annexin VIII cDNA cloning

Using the amino acid sequence information from annexin V peptide P30/I [3] for the synthesis of oligonucleotides, subsequent screening of a placental library revealed the existence of additional clones representing a new annexin — annexin VIII [14]. The corresponding mRNA defines a 1086 nt long open reading frame with the coding potential of 326 amino acids and a calculated molecular mass of 36706 Da. The context of the first AUG differs considerably from Kozak consensus sequences (AGAGAUGG) and there is no in-frame stop codon helping to definitively assign the translation start site to this AUG. However, consistent with the indicated translational start is that the second encoded amino acid is an alanine. All human annexins except annexin II start with an alanine. The 3′-non-translated region is 852 bases long and contains, in addition to the consensus polyadenylation signal 22 bases upstream of the poly(A) stretch, another AATAAA motif 240 bases 5′ to the functional polyadenylation site. In none of the isolated cDNAs was this site used as the polyadenylation signal despite the occurrence of a sequence TGTGTTAT 31 bases downstream, which is homologous to the consensus sequence YGTGTTYY found 3′ to the polyadenylation site of many eukaryotic class II genes. Northern blot analysis of placental mRNA detects a single hybridizing species of 2100–2200 bases and indicates about 20 times less annexin VIII than annexin V mRNA in this tissue.

Compared to annexin V annexin VIII is a total of five amino acids longer: the *N*-terminal segment is six amino acids longer containing a clustered set of three tryptophans whereas the linker connecting the second and third repeat is one amino acid shorter. The highest degree of identity exists with annexin V (55.9 %), annexin IV (55.8 %) and the first half of annexin VI (52.1 %). All other annexins show less than 50 % identity. Annexin VIII contains a potential N-glycosylation site (Asn–Lys–Ser) in the third repeat and four cysteines. Annexin VIII also shows a surplus of seven acidic amino acids resulting in an isoelectric point between 5.2 and 5.8 with a maximum of 5.45 as measured with the recombinant protein. The reason for this heterogeneity was not determined but could be because of different extents of cysteine oxidation and/or alternative disulphide formation of the protein as it is produced in bacteria.

## Annexin V and VIII genes

Southern blot analysis using annexin V and annexin VIII cDNA as probes showed the existence of one gene of moderate size for each of these annexins [3,14]. Therefore, no pseudogenes should be expected as with annexin II [23], for which one functional and three pseudogenes have been isolated.

Evidence is now accumulating that the individual human genes of the annexin supergene family are located on different chromosomes. *In situ* hybridization and segregation analysis in somatic cell hybrids maps the genes for annexin I (*anx1*) to chromosome 9q11→q22 [24], annexin II (*anx2*) to chromosome 15q21-q22 with the three pseudogenes located at 4q21-q31.1, 9pter-q34 and 10cen→q24 [23,24], annexin III (*anx3*) to 4q13-23 [25] and annexin VI (*anx6*) to chromosome 5q32-q34 [26]. The annexin V (*anx5*) gene was mapped to chromosome 4q28→32 [27] or 4q26→28 [28]. Using a very sensitive non-isotopic *in situ* hybridization procedure [29] we could determine the map position of *anx5* as 4q27 and the annexin VIII (*anx8*) gene even more precisely to chromosome 10q11.2 (P. F. Ambros *et al.*, unpublished work).

Up to now the gene structures of two human (*anx2* [23] and *anx1* [30]) and three animal (rat *anx1* [30], avian *anx1* [31] and mouse *anx1* [32]) annexins have been published. These annexins are encoded by genes containing 13 exons. The number of introns and the sites where introns interrupt the coding sequences are conserved not only between one gene and the same annexin gene in different species but also between different annexins. The organizations of the human annexin V and VIII genes are further examples of this structure (P. F. Ambros *et al.*, unpublished work). The most striking feature is that the exon–intron distribution does not mirror the fourfold repetitive nature of annexins [23,30–32] (P. F. Ambros *et al.*, unpublished work).

## The phospholipid-binding properties

The paradigm of annexin V is centred around its phospholipid-binding property. This activity forms the basis of the mechanisms of its *in vitro* observed biological actions, like the anticoagulant and anti-phospholipase activity, and has been studied intensively. Annexin V is a water-soluble protein that does not bind to phospholipid membranes in the absence of $Ca^{2+}$, although it can develop a high affinity for such membranes in the presence of $Ca^{2+}$. The mechanism of binding is not yet fully understood but some insights have been obtained by the research of many laboratories.

Small amounts of $Ca^{2+}$ (10–30 μM) induce a conformational change of the annexin V molecule, creating and/or exposing sites that allow and/or mediate the interaction with phospholipids [18] (C. P. M. Reutelingsperger, unpublished work). There is evidence that the molecule undergoes a series of conformational changes with increasing $Ca^{2+}$ concentration and each of these conformations probably bears a certain set of phospholipid-interacting sites (Fig. 1 and see below).

The $Ca^{2+}$ requirement of annexin V for binding to phospholipids clearly depends on the composition of the phospholipid membrane [33,34]. For binding to a phosphatidylcholine membrane annexin V requires $Ca^{2+}$ concentrations between 10 and 100 mM. Insertion of sphingomyelin into a phosphatidylcholine membrane evokes a slight decrease in $Ca^{2+}$ requirement, whereas insertion of phosphatidylserine and cardiolipin causes a considerable drop in the $Ca^{2+}$-sensitivity to 10–40 μM.

Once bound to the surface the protein may undergo another conformational change at low $Ca^{2+}$ concentrations that is catalysed by the phospholipids [18]. On the basis of work by Huber *et al.* [35] and Mosser *et al.* [36], Brisson *et al.* [37] reported the absence of conformational differences between the soluble and membrane-bound form of annexin V (see chapter 10 by Huber *et al.*). This was observed at high $Ca^{2+}$ concentrations, which does not exclude phospholipid-induced conformational changes under submillimolar $Ca^{2+}$ conditions. Phospholipid-bound annexin V occurs as a slightly elongated protein with dimensions of $64 \times 40 \times 30$ Å [35,36] covering approximately 25 $nm^2$ of phospholipid surface [33,36]. The binding is most likely not accompanied by insertion of protein domains into the bilayer and is of a typical extrinsic nature [33,35]. The association with the membrane is rapid and reversible upon removal of the $Ca^{2+}$ ions. A cycle of binding and release provoked by a sequential increase and decrease of the $Ca^{2+}$ concentration can be repeated several times without affecting the binding characteristics of annexin V and the integrity of the phospholipid membrane [3,34,38].

Following binding, a process of protein polymerization proceeds at the phospholipid surface. This two-dimensional polymerization occurs by non-covalent and reversible protein–protein interactions [33] (H. A. M. Andree *et al.*, unpublished work) resulting in clusters of three molecules forming triskelion structures [36]. The resolved subdomain structures of lipid-bound annexin V reflects the domain organization of its linear amino acid sequence and the three-dimensional folding of the polypeptide chain in the presence of 1.6 mM $Ca^{2+}$ [35]. This group also elucidated the presence of three high-affinity $Ca^{2+}$ binding sites in domains I, II and IV [39], that employ the structural motif (M,L)–KG–(A,L)–GT, which is present in these domains. The $Ca^{2+}$ ion uses this motif to occupy six of its coordination sites leaving the seventh site for interaction with a phosphoryl moiety of the membrane [39]. Although not yet unambiguously demonstrated, this $Ca^{2+}$-bridging model of interaction is supported by several other observations and considerations. The elucidated three-dimensional structure shows a folding of the molecule with a convex and a concave site. The $Ca^{2+}$-binding sites are located at the convex site and circumstantial evidence points to an orientation of this site in juxtaposition to the phospholipid membrane [35–37,40].

**Fig. 1. $Ca^{2+}$-induced conformational changes of annexin V.**

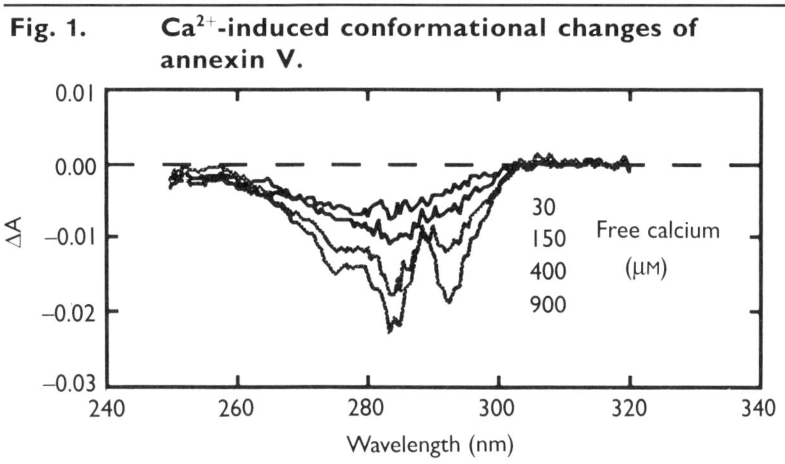


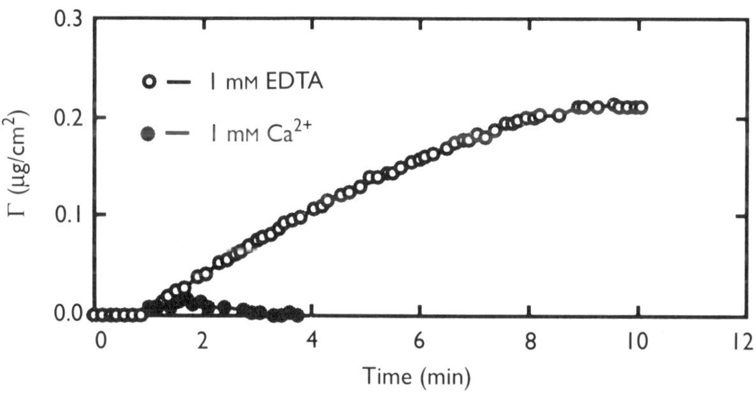


*Upper panel: Difference spectra of annexin V between conformations in the presence of free $Ca^{2+}$ and conformations in the presence of EGTA. The spectrum of 1 mg annexin V/ml of 20 mM Hepes and 30 μM EGTA, pH 7.4, was recorded by a Uvikon 860. $Ca^{2+}$ was then titrated, resulting in the indicated free $Ca^{2+}$ concentrations. At each concentration a spectrum was recorded from which the initial spectrum was subtracted.*

*Lower panel: Thermostability of annexin V in the presence of EDTA. Samples of 20 μg annexin V/ml of 50 mM Tris/HCl pH 7.8, 100 mM NaCl was incubated either in the presence of 1 mM EDTA or 1 mM $Ca^{2+}$ for 10 min at 56 °C. The resulting annexin V preparations were then assayed for their phospholipid-binding activity by ellipsometry, as described elsewhere (H. A. M. Andree et al. unpublished work).*

The $Ca^{2+}$-bridging model probably does not explain fully the observed binding characteristics of annexin V. Additionally, the association of annexin with the phospholipid membrane may be governed by ionic and hydrophobic forces [38,41] (H. A. M. Andree, unpublished work) that mediate protein–phospholipid and protein–protein interactions [42].

Thus, the binding of annexin V to phospholipid membranes involves a complex concert of different types of interactions, making it difficult or even impossible to determine a binding parameter like the $K_d$. The affinity of annexin V for phosphatidylserine-containing membranes is high, which at least partly explains the observed biological activities.

The on/off switches of these sites are regulated by $Ca^{2+}$ concentrations and the phospholipid composition of the membranes, and possibly even by other proteins. Insights herein would provide understanding of the physiological importance of annexin V. On the basis of structural considerations these binding characteristics of annexin V may be extrapolated to the other members of the annexin family [35,43].

## *In vitro* biological activities and consequences for understanding of physiological roles

It has been impossible to speculate about the physiological function of annexin VIII so far because no protein product could be detected in human placenta and various other tissues (C. P. M. Reutelingsperger, unpublished work). The discovery of annexin VIII, by screening a placental cDNA library with a probe for annexin V, was incidental [14]. Despite the presence of small amounts of annexin VIII mRNA in placenta, the protein product has remained elusive, casting some doubts on the existence of an annexin VIII protein. Recent publications by the group of Tsao [44,45] describe the identification of an annexin VIII protein in rabbit lung. The partial sequence of the described $Ca^{2+}$-dependent phospholipid-binding protein (33000 PLBP) shows a 56 % identity to the corresponding amino acid sequence of human annexin V. The complete comparison of this sequence with all eight human annexins reveals an 88 % identity of the 33000 PLBP with the first repeat of annexin VIII. Identities to corresponding segments in other annexins vary between 36 and 58 % (Table 1). On the basis of the high interspecies conservation of annexins, the 33000 PLBP is likely to be the rabbit counterpart of human annexin VIII. Preliminary results in our laboratory indicate the presence of annexin VIII in human lung but not in human liver, brain, kidney and spleen. Although more data are needed, it appears that annexin VIII is the least

**Table 1. Alignment of the rabbit 33 kDa phospholipid-binding protein (PLBP) peptide with the corresponding segments of the first repeat of each of the human annexins**

| | Sequence | Identity (%) |
|---|---|---|
| PLBP | MKGIGTNEQAIIDVLTRRSSAQRQQIAKSFKAQFGSDLTED | |
| Annexin I | IMVK.VD.AT...I..K.NN......KAAYLQET.KP.D.T | 41.5 |
| Annexin II | I.TK.VD.VT.VNI..N..N....D..FAYQRRTKKE.ASA | 36.6 |
| Annexin III | IR....D.KML.SI..E..N....L.V.EYQ.AY.KE.KD. | 48.8 |
| Annexin IV | ...L..D.D...S..AY.NT....E.RTAY.STI.R..ID. | 53.7 |
| Annexin V | ...L..D.ES.LTL..S..N....E.SAA..TL..R..LD. | 56.1 |
| Annexin VI/1 | ...F.SDKE..L.II.S..NR...EVCQ.Y.SLY.K..IA. | 46.3 |
| Annexin VI/2 | ...L..D.DT...II.H..NV.....RQT..SH..R..MT. | 58.5 |
| Annexin VII | ...F..D....V..VAN..ND...K.KAA..TSY.K..IK. | 56.1 |
| Annexin VIII | ................K..NT..............K....T | 87.8 |

*Dots indicate the same amino acid as in the rabbit peptide. Percent identities between the peptide and the different annexins are given. Annexin VI is split into two halves and the rabbit peptide compared to the first and fifth repeat.*

ubiquitous protein of the annexin family and, thus, possibly serves a unique and tissue-specific function. This should stimulate future research to unravel the physiological function of annexin VIII.

In contrast to annexin VIII, annexin V is one of the most widely spread annexins and is present in large quantities in most tissues

**Fig. 2. Indirect immunofluorescent staining of annexin V in HeLa cells.**

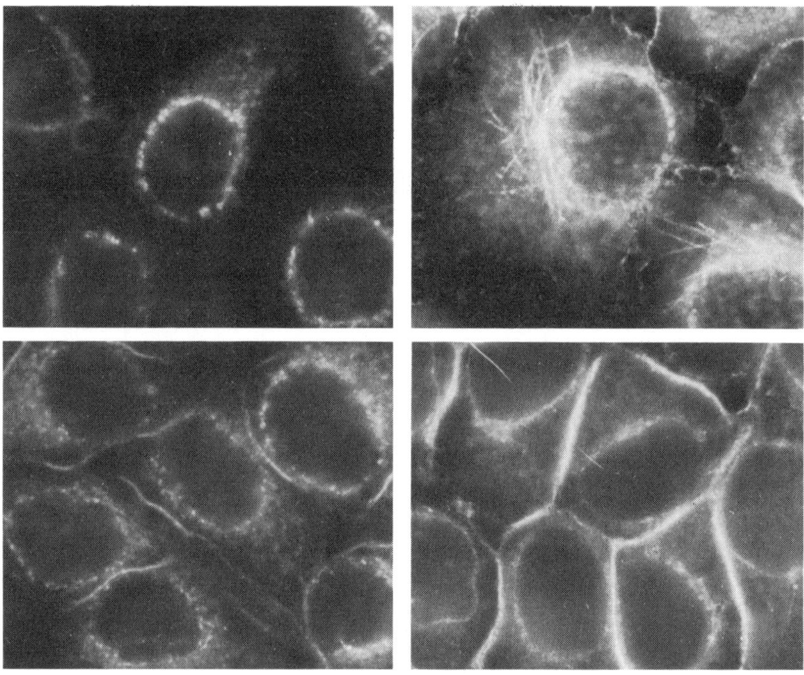

*HeLa cells were plated at low density on glass coverslips and allowed to grow to confluence. At different stages the cells were fixed, permeabilized and incubated with annexin V-specific rabbit antibodies and fluorescein-conjugated anti-rabbit IgG. The upper left panel shows HeLa cells directly following seeding. Upper right, lower left and lower right show HeLa cells through various stages until reaching confluency.*

[2,9,46–54]: it is localized intracellularly in most cell types in the cytoplasm and is associated with intracellular vesicles and plasma membranes (Fig. 2). Extracellular localization associated with cells [22] or extracellular matrix [55], or in body fluids like blood plasma, amniotic fluid and seminal plasma, has been reported [56,57]. It remains unknown whether the extracellular annexin V originates from selective externalization processes or simply from leakage out of damaged cells.

Several *in vitro* biological activities have been assigned to annexin V, including anticoagulant [1,2,4,58–65] and anti-phospholipase activities [14,47,65] which are dealt with by most papers. The basis for its inhibitory capacity lies in its potency to bind with high affinity to the phospholipid surface and to compete for binding with other proteins without affecting their enzymatic activity *per se*. As the phospholipid-binding characteristic in essence is shared by all annexins, it is not surprising that the anticoagulant and anti-phospholipase activity appears to be a common feature of the annexins; however, the anticoagulant efficacy is different. Annexin V appears to be one of the most potent inhibitors of coagulation [65,66].

The proposed substrate-depletion model, which explains the mechanism by which all the annexins exhibit anti-phospholipase activity, would argue against a physiologically important function as a phospholipase inhibitor and analogously as an inhibitor of coagulation [5,67]. However, the teleological difference between phospholipids in the context of phospholipase activity and phospholipids in the coagulation process necessitates some caution about drawing analogous conclusions on that basis.

The low concentration of annexin V found in the blood circulation does not favour an important function as an anticoagulant [56]. On the other hand the high concentration present in endothelial cells lining the lumen of blood vessels would have an important contribution in lowering locally the thrombogenicity once the cells are damaged [56,68].

It is now generally believed that annexin V has an intracellular physiological function that relates to its phospholipid-binding properties. A role in exocytosis, as suggested for some other annexins, is unlikely because annexin V lacks a vesicle-aggregating activity [34]. A not yet clearly defined role of annexin V in cell cycle regulation has been suggested [69]. Annexin V has also been proposed to play a role in $Ca^{2+}$ transport mechanisms through the cytoplasm or across membranes [70]. The high number of copies of annexin V in endothelial cells and fibroblasts, however, argues against a function as a $Ca^{2+}$ channel [70].

## Conclusion

On collaborating the available data, one is left with a vague picture of the true physiological function of annexin V. The behaviour of isolated purified proteins *in vitro* may or may not reflect the significance of these proteins in physiology, a classic dilemma often faced by scientists. The short history of the annexins is a pertinent illustration of such dilemma. The number of possible physiological functions of the annexins equals or even exceeds the number of their discovered *in vitro* activities. Evidence of function might be found in the pathogenesis provoked by an annexin deficiency. Up to date no such pathogenic deficiencies have been described.

## References


1. Reutelingsperger, C. P. M., Hornstra, G. & Hemker, H. C. (1985) Eur. J. Biochem. **151**, 625–629
2. Reutelingsperger, C. P. M., Kop, J. M. M., Hornstra, G. & Hemker, H. C. (1988) Eur. J. Biochem. **173**, 171–178
3. Maurer-Fogy, I., Reutelingsperger, C. P. M., Pieters, J., Bodo, G., Stratowa, C. & Hauptmann, R. (1988) Eur. J. Biochem. **174**, 585–592
4. Iwasaki, A., Suda, M., Nakao, H., Nagoya, T., Saino, Y., Arai, K., Mizoguchi, T., Sato, F., Yoshizaki, T., Hirata, M., Miyata, T., Shidara, Y., Murata, M. & Maki, M. (1987) J. Biochem. (Tokyo) **102**, 1261–1273
5. Kaplan, R., Jaye, M., Burgess, W. H., Schlaepfer, D. D. & Haigler, H. T. (1988) J. Biol. Chem. **263**, 8037–8043
6. Grundmann, U., Abel, K.-J., Bohn, H., Löbermann, H., Lottspeich, F. & Küpper, H. (1988) Proc. Natl. Acad. Sci. U.S.A. **85**, 3708–3712
7. Funakoshi, T., Hendrickson, L. E., McMullen, B. A. & Fujikawa, K. (1987) Biochemistry **26**, 8087–8092
8. Pepinsky, R. B., Tizard, R., Mattaliano, R. J., Sinclair, L. K., Miller, G. T., Browning, J. L., Chow, E. P., Burne, C., Huang, H.-S., Pratt, D., Wachter, L., Hession, C., Frey, A. Z. & Wallner, B. P. (1988) J. Biol. Chem. **263**, 10799–10811
9. Kaetzel, M. A., Hazarika, P. & Dedman, J. R. (1989) J. Biol. Chem. **264**, 14463–14470
10. Fernández, M. P., Selmin, O., Martin, G. R., Yamada, Y., Pfäffle, M., Deutzmann, R., Mollenhauer, J. & von der Mark, K. (1990) J. Biol. Chem. **265**, 8344
11. Geisow, M. J. & Walker, J. H. (1986) Trends Biochem. Sci. **11**, 420–423
12. Geisow, M. J., Fritsche, V., Hexham, J. M., Dash, B. & Johnson, T. (1986) Nature (London) **320**, 636–638
13. Crumpton, M. J. & Dedman, J. R. (1990) Nature (London) **345**, 212
14. Hauptmann, R., Maurer-Fogy, I., Krystek, E., Bodo, G., Andree, H. A. M. & Reutelingsperger, C. P. M. (1989) Eur. J. Biochem. **185**, 63–71
15. Kozak, M. (1984) Nature (London) **308**, 241–246
16. Bohn, H. (1985) Behring Inst. Mitt. **78**, 70–82
17. Kretsinger, R. H. & Creutz, C. E. (1986) Nature (London) **320**, 573
18. Schlaepfer, D. D., Mehlman, T., Burgess, W. H. & Haigler, H. T. (1987) Proc. Natl. Acad. Sci. U.S.A. **84**, 6078–6082
19. Burgoyne, R. D. (1987) Trends Biochem. Sci. **12**, 85–86
20. Fernández, M. P., Selmin, O., Martin, G. R., Yamada, Y., Pfäffle, M., Deutzmann, R., Mollenhauer, J. & von der Mark, K. (1988) J. Biol. Chem. **263**, 5921–5925
21. Mollenhauer, J., Bee, J. A., Lizarbe, M. A. & von der Mark, K. (1984) J. Cell. Biol. **98**, 1572–1578
22. Pfäffle, M., Ruggiero, F., Hofmann, H., Fernández, M. P., Selmin, O., Yamada, Y., Garrone, R. & von der Mark, K. (1988) EMBO J. **7**, 2335–2342
23. Spano, F., Raugei, G., Palla, E., Colella, C. & Melli, M. (1990) Gene **95**, 243–251
24. Huebner, K., Cannizzaro, L. A., Frey, A. Z., Hecht, B. K., Hecht, F., Croce, C. M. & Wallner, B. P. (1988) Oncogene Res. **2**, 299–310
25. Tait, J. F., Frankenberry, D. A., Miao, C. M., Killary, A. M., Adler, D. A. & Disteche, C. M. (1991) Genomics **10**, 441–448
26. Davies, A. A., Moss, S. E., Crompton, M. R., Jones, T. A., Spurr, N. K., Sheer, D., Kozak, C. & Crumpton, M. J. (1989) Hum. Genet. **82**, 234–238
27. Modi, W. S., Seuànez, H. N., Jaye, M., Haigler, H. J., Kaplan, R. & O'Brien, S. J. (1989) Cytogenet. Cell Genet. **52**, 167–169

28. Tait, J. F., Frankenberry, D. A., Shiang, R., Murray, J. C., Adler, D. A. & Disteche, C. M. (1991) Cytogenet. Cell Genet. **57**, 187–192
29. Ambros, P. F. & Karlic, H. I. (1987) Hum. Genet. **77**, 251–254
30. Kovacic, R. T., Tizard, R., Cate, R. L., Frey, A. Z. & Wallner, B. P. (1991) Biochemistry **30**, 9015–9021
31. Hitti, Y. S. & Horseman, N. D. (1991) Gene **103**, 185–192
32. Horlick, K. R., Cheng, I. C., Wong, W. T., Wakeland, E. K. & Nick, H. S. (1991) Genomics **10**, 365–374
33. Andree, H. A. M., Reutelingsperger, C. P. M., Hauptmann, R., Hemker, H. C., Hermens, W. T. & Willems, G. (1990) J. Biol. Chem. **265**, 4923–4928
34. Blackwood, R. A. & Ernst, J. D. (1990) Biochem. J. **266**, 195–200
35. Huber, R., Römisch, J. & Paques, E.-P. (1990) EMBO J. **9**, 3867–3874
36. Mosser, G., Ravanat, C., Freyssinet, J.-M. & Brisson, A. (1991) J. Mol. Biol. **217**, 241–245
37. Brisson, A., Mosser, G. & Huber, R. (1991) J. Mol. Biol. **220**, 199–203
38. Meers, P., Daleke, D., Hong, K. & Papahadjopoulos, D. (1991) Biochemistry **30**, 2903–2908
39. Huber, R., Schneider, M., Mayr, I., Römisch, J. & Paques, E.-P. (1990) FEBS Lett. **275**, 15–21
40. Meers, P. (1990) Biochemistry **29**, 3325–3330
41. Tait, J., Gibson, D. & Fujikawa, K. (1989) J. Biol. Chem. **264**, 7944–7949
42. Ahn, N. G., Teller, D. C., Bienkowski, M. J., McMullen, B. A., Lipkin, E. W. & de Haën, C. (1988) J. Biol. Chem. **263**, 18657–18663
43. Barton, G. J., Newman, R. H., Freemont, P. S. & Crumpton, M. J. (1991) Eur. J. Biochem. **198**, 749–760
44. Tsao, F. H. (1990) Biochim. Biophys. Acta **1045**, 29–39
45. Tsao, F. H., Hull, W. M., Strickland, M. S., Whitsen, J. A., Foo, T. K., Zografi, G. & DeLuca, P. M. (1991) Biochim. Biophys. Acta **1081**, 141–150
46. Funakoshi, T., Heimark, R. L., Hendrickson, L. E., McMullen, B. A. & Fujikawa, K. (1987) Biochemistry **26**, 5572–5578
47. Haigler, H. T., Schlaepfer, D. D. & Burgess, W. H. (1987) J. Biol. Chem. **262**, 6921–6930
48. Kanna, N. C., Helwig, E. D., Ikebuchi, N. W., Fitzpatrick, S., Bajwa, R. & Waisman, D. M. (1990) Biochemistry **29**, 1852–1862
49. Römisch, J. & Heimburger, N. (1990) Biol. Chem. Hoppe Seyler **371**, 383–388
50. Pula, G., Bianchi, R., Ceccarelli, P., Giambanco, I. & Donato, R. (1990) FEBS Lett. **277**, 53–58
51. Kobayashi, R. & Tashima, Y. (1989) Biochem. Biophys. Res. Commun. **162**, 15–23
52. Kobayashi, R., Nakayama, R., Ohta, A., Sakai, S. F., Sakuragi, S. & Tashima, Y. (1990) Biochem. J. **266**, 505–511
53. Woolgar, J. A., Boustead, C. M. & Walker, J. H. (1990) J. Neurochem. **54**, 62–71
54. Bolton, C., Elderfield, A.-J. & Flower, R. J. (1990) J. Neurochem. **29**, 173–181
55. Genge, B. R., Wu, L. N. Y., Adkinson, H. D. & Wuthier, R. E. (1991) J. Biol. Chem. **266**, 10678–10658
56. Flaherty, M. J., West, S., Heimark, R. L., Fujikawa, K. & Tait, J. F. (1990) J. Lab. Clin. Med. 174–181
57. Christmas, P., Callawy, J., Falon, J., Jones, J. & Haigler, H. T. (1991) J. Biol. Chem. **266**, 2499–2507
58. Gramzinski, R. A., Broze, G. J. & Carson, S. D. (1989) Blood **73**, 983–989
59. Kondo, S., Noguchi, M., Funakoshi, T. & Kisiel, W. (1987) Thromb. Res. **48**, 449–459
60. Thiajagaran, P. & Tait, J. F. (1990) J. Biol. Chem. **265**, 17420–17423
61. Van Heerde, W. I., Poort, S., van't Veer, C., Reutelingsperger, C. P. M. & de Groot, P. G. (1991) Thromb. Haemostas. **65**, 1310
62. Yoshizaki, H., Tanabe, S., Yokoyama, T., Murakami, A., Wada, Y., Miyamoto, S. & Maki, M. (1991) Thromb. Haemostas. **65**, 1208
63. Ravanat, C., Archipoff, G., Freyssinet, J.-M., Schuhler, S., Beretz, A. & Cazenave, J.-P. (1991) Thromb. Haemostas. **65**, 776

64. Römisch, J., Schorlemmer, U., Fickenscher, K., Paques, E.-P. & Heimburger, N. (1990) Thromb. Res. **60**, 355–366
65. Tait, J. F., Sakata, M., McMullen, B. A., Miao, C. H., Funakoshi, T., Hendrickson, L. E. & Fujikawa, K. (1988) Biochemistry **27**, 6268–6276
66. Chap, H., Comfurius, P., Bevers, E. M., Fauvel, J., Vicendo, P., Douste-Blazy, L. & Zwaal, R. F. A. (1988) Biochem. Biophys. Res. Commun. **150**, 972–978
67. Davidson, F. F., Dennis, E. A., Powell, M. & Glenney, J. R., Jr. (1987) J. Biol. Chem. **262**, 1698–1707
68. van Gool, R., Andree, H. A. M., Hemker, H. C. & Reutelingsperger, C. P. M. (1990) in Trends in Haemostasis. Coagulation protein, Endothelium and Tissue Factors (Arenberg, J. H., Heene, D. L. & Schettler, G., eds.), pp. 136–151, Springer Verlag, Heidelberg
69. Schlaepfer, D. D. & Haigler, H. T. (1990) J. Clin. Invest. **111**, 229–238
70. Rojas, E., Pollard, H. B., Haigler, H. T., Parra, C. & Burns, A. L. (1990) J. Biol. Chem. **265**, 21207–21215

# Placental annexins: isolation, characterization and possible functions

**Masahiro Maki, Yoshihiro Shidara and Hirokazu Sato**

Department of Obstetrics and Gynecology, Akita University School of Medicine, 1-1-1 Hondo, Akita, 010, Japan

## Introduction

The placental circulation is composed of two systems: maternal intervillous and fetal intravillous. Although the fetal circulation has its specific bypasses, the basic system of blood vessels i.e. a composition of artery-arteriole-capillary-venule-vein, is the same as in the general circulation. The maternal intervillous circulation is, however, different from the general circulation system for a number of reasons: (i) the intervillous space is structurally different from the normal vascular anatomy, there being no endothelial cells lining the luminal surface of the intervillous space, (ii) the blood flow in the intervillous space is slow; (iii) maternal blood during pregnancy is hypercoagulable and hypofibrinolytic; and (iv) the placenta is rich in tissue-factor- and plasminogen-activator inhibitor. These conditions fit very well with Virchow's theory of thrombogenesis. In fact, thrombosis is frequently observed in the intervillous space in some pathological conditions such as pre-eclampsia-eclampsia and systemic lupus erythematosis. However, the intervillous circulation is normally well maintained to ensure fetal well-being. It is concluded, therefore, that antithrombotic mechanisms function in this tissue.

Anti-thrombotic factors already known to be located on the surface of the chorionic villi include platelet-aggregation inhibitors such as prostacyclin-like substance [1], ADPase which blocks ADP-induced platelet aggregation [2], thrombomodulin [3], tissue plasminogen activator [4] and a heparin-like substance demonstrated by toluidine blue metachromasia.

We previously isolated two placental anti-thrombotic proteins [5]. One possessed an anticoagulant activity and the other had an inhibitory activity on platelet aggregation. The amino acid sequence of this anticoagulant protein was deduced from cDNA cloning and protein sequencing [6–8]. Because this protein binds $Ca^{2+}$ and phospholipid, Maki *et al.* [9] designated it calphobindin (calcium phospholipid binding protein; CPB). Now the family proteins are called annexins. In this chapter, we review our data on the purification, characterization and possible functions of placental annexins.

We have isolated and identified three potent inhitors of the activation process of Factor X and prothrombin (CPBs I, II and III) from human placenta, which function in a manner dependent on their binding to $Ca^{2+}$-phospholipid. Binding of Factor X or prothrombin to phospholipid vesicles in the presence of $Ca^{2+}$ was inhibited in a dose-dependent fashion by CPBs I, II and III. Partial amino acid sequencing of the inhibitors revealed that they were annexin family proteins, corresponding to annexins V, VI and III, respectively. The inhibitors were expressed not only in placenta, but also in other various organs and in body fluids, especially cavity fluids. They may play a role in antithrombotic defense mechanisms in the intervillous circulation and also in the general circulation when endothelial cells are damaged.

## Purification and characterization of placental calphobindins (annexins)

On discovering that the placental coagulation inhibitor (CPB I) was a member of the annexin family, we modified our original isolation procedure by employing a $Ca^{2+}$-chelating agent as shown in Figs. 1 and 2.

Five frozen placentae were thawed and homogenized after the fetal membranes and umbilical cord had been removed. The preparation procedures were all carried out at 4 °C. The EDTA extract was treated by fractional ammonium sulphate precipitation, and the precipitated proteins (80% saturation) were purified by several steps of column chromatography. The flow-through fraction and the eluates corresponding to 0.10 M, 0.18 M and 0.20 M NaCl in the linear gradient elution of the Q-Sepharose column contained lipocortin II (annexin II), CPB III (annexin III), CPB I (annexin V) and CPB II (annexin VI), respectively. Each fraction was further purified to homogeneity, as judged by migration as a single protein in SDS/PAGE. The yields of CPBs I, II and III from one placenta were approximately 30, 3 and 1 mg, respectively. There are at least six placental annexin family proteins (annexins I, II, III, IV, V and VI), although we failed to isolate annexins I and IV in these procedures.

### Characterization of placental calphobindins

The complete amino acid sequences of CPBs I and II deduced from cDNA cloning [6,11,12] were identical to those of annexins V and VI, respectively.

Partial sequence analysis of native CPB III revealed that it was the same as annexin III.

Two-dimensional electrophoresis of annexins II, III, V and VI purified from placental extract by adsorption on phospholipid vesicles in

**Fig. 1. Preparation procedure of placental annexins (early steps).**

Fresh human placenta (2.5 kg)
↓
Homogenized in 5 l of 50 mM Tris-HCl with 5 mM $CaCl_2$ (pH 7.4)
↓ centrifuged for 15 min at 10000 *g*
Rehomogenized in the same buffer
↓ centrifuged for 15 min at 10000 *g*
Homogenized in 2.5 l of 50 mM Tris-HCl with 5 mM EDTA (pH 7.4)
↓ centrifuged for 15 min at 10000 *g*
Supernatant: 30 % of $(NH_4)_2SO_4$ saturation
↓ centrifuged for 30 min at 10000 *g*
Supernatant: 80 % of $(NH_4)_2SO_4$ saturation
↓ centrifuged for 30 min at 10000 *g*
Precipitate: dissolved in 0.5 l of 50 mM Tris-HCl (pH 7.4)
↓ dialysed against 50 mM Tris-HCl (pH 7.4)
Q-Sepharose column (5 × 7.6 cm)
↓ eluted with linear NaCl gradient (0–0.3 M)
Active fractions
(1) 0.10 M NaCl: annexin III
(2) 0.18 M NaCl: annexin V
(3) 0.20 M NaCl: annexin VI

the presence of $Ca^{2+}$ revealed pI values in the range 4.9–8.1. Molecular masses of most of the annexins were around 35 kDa, whilst that of annexin VI was approximately 70 kDa [13].

## Anticoagulant activity

CPBs prolonged the prothrombin time dose dependently and activated partial thromboplastin time as shown in Fig. 3. The inhibitory activity of CPB II was strongest, followed by CPBs I and III (Fig. 3). They did not inhibit Factor Xa or thrombin. Factor X activation was analysed through the intrinsic pathway by reaction with Factor VIII, Xa and phospholipid-$Ca^{2+}$ in the presence of various concentrations of CPB I. Formation of Xa was measured using the chromogenic synthetic substrate S-2337. Xa formation was inhibited dose-dependently with increasing amounts of

CPB I (Fig. 4). CPB I also inhibited the activation process of Factor X through the extrinsic pathway. Factor X activation with Russell's viper venom, where phospholipid is unnecessary, was not inhibited by CPB I (Fig. 5).

**Fig. 2. Preparation procedure of placental annexin (late steps).**

Active fractions

(1) Annexin III fraction
↓ concentrated with Diaflo YM 10
Sephadex G-100 column (2 × 90 cm)
↓ dialysed against 25 mM acetate buffer (pH 5.2)
Mono S column (fast protein liquid chromatography system)
↓ eluted with linear NaCl gradient (0–0.4 M)
Pure fraction (0.24 M NaCl: annexin III)
(2) Annexin V fraction
↓ dialysed against 20 mM phosphate buffer (pH 7.0)
Blue sepharose CL-6B column (3 × 15 cm)
↓ concentrated and dialysed against 20 mM Tris-HCl (pH 7.4)
Sephadex G-100 column (2 × 90 cm)
↓
Pure fraction (annexin V)
(3) Annexin VI fraction
↓ dialysed against 20 mM phosphate buffer (pH 7.0)
Blue sepharose CL-6B column (3 × 15 cm)
↓ concentrated and dialysed against 20 mM Tris-HCl (pH 8.5)
Mono Q column (fast protein liquid chromatography system)
↓ eluted with linear NaCl gradient (0–0.5 M) and concentrated Sephadex G-100 column (2 × 90 cm)
↓
Pure fraction (annexin VI)

In the same way, prothrombin was activated by incubating a mixture of prothrombin, Xa, phospholipid and $Ca^{2+}$ in the presence of various concentrations of CPB I. The generation of thrombin was measured by the synthetic substrate method using RM-601-2. Prothrombin activation was inhibited dose-dependently by CPB I. The $Ca^{2+}$-phospholipid independent activation of prothrombin by Ecarin was not inhibited by CPB I (Fig. 6). Binding of prothrombin to phospholipid vesicles in the presence of $Ca^{2+}$ was also inhibited dose-dependently by increasing amounts of CPB I (Fig. 7). In a similar manner, binding of Factor X to phospholipid vesicles was inhibited by CPB I [12].

### Phospholipid-binding

Phospholipids were extracted from 1 g of placental tissue thromboplastin (Thromorel) by 20 ml of an organic solvent mixture consisting of 2 vol. of heptane and 1 vol. of butanol. The phospholipid extract (0.77 mg) was

**Fig. 3. Effects of CPBs I (annexin V), II (annexin VI) and III (annexin III) on prothrombin time (a) and activated partial thromboplastin time (b).**

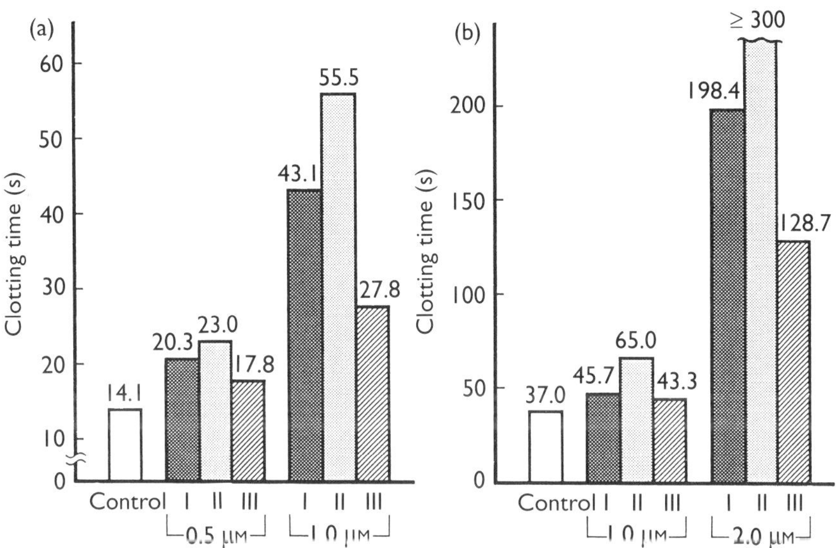


*PT was measured using normal standard plasma (100 μl) and Lyoplastin-$CaCl_2$ (150 μl) in the presence or absence of CPBs (100 μl). APTT was measured using normal standard plasma (50 μl) and Cephotest (50 μl)–0.02 M $CaCl_2$ (100 μl) in the presence of various concentrations of CPBs (100 μl).*

poured into a glass tube (2 × 10.5 cm) and coated onto the internal wall of the tube by evaporating the organic solvent. A mixture of 75 μl of 250 μg/ml CPB I and 75 μl of various concentrations of $CaCl_2$ (0–1000 mmol) was incubated in the phospholipid-coated glass tube at 37 °C for 5 min, and the CPB I concentration in the supernatant was measured by e.l.i.s.a. There was no change in the concentration of CPB I in the absence of $CaCl_2$. When $CaCl_2$ solution was added to the tube, the CPB I concentration in the supernatant decreased and eventually disappeared at high concentrations of $CaCl_2$. This demonstrates that CPB I binds to the phospholipid on the glass tube wall in the presence of $CaCl_2$. The CPB I bound to the phospholipid was eluted into solution by the

**Fig. 4. Effect of CPB I on Factor X activation by intrinsic pathway.**

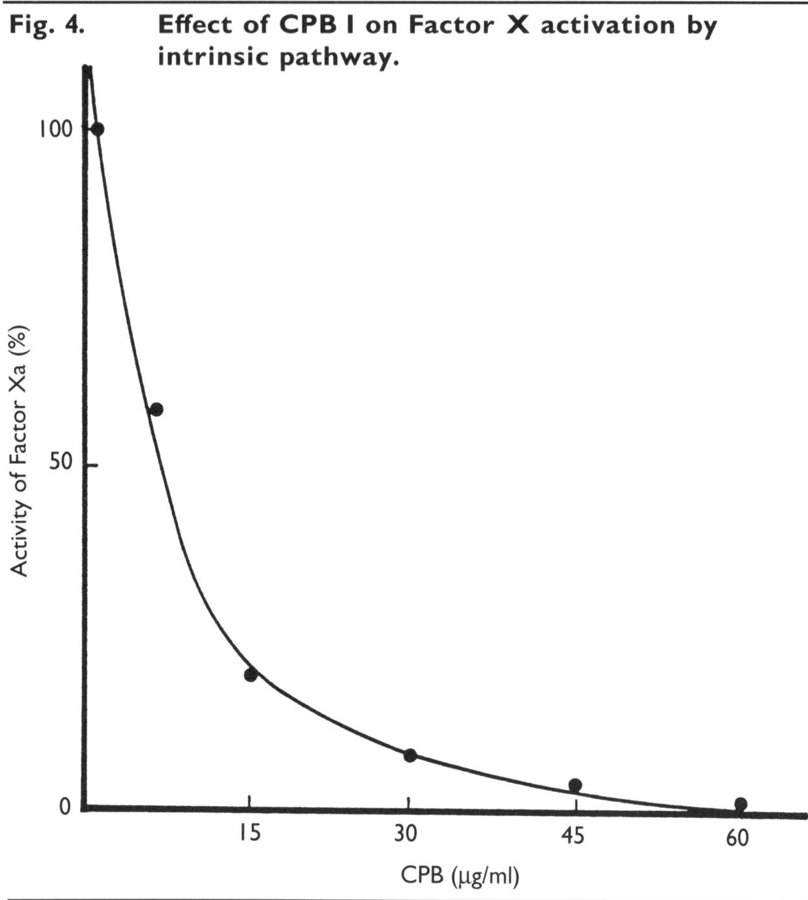


*A reaction mixture of 200 μl of a mixture of Factors IXa and X from a kit of Factor VIII assay, 100 μl of 1.2% Factor VIII, 100 μl of 0.1 M $CaCl_2$, 20 μl of 31.8 μg/ml phospholipid, and 20 μl of various concentrations of CPB I (0–60 μg/ml) was incubated at 37 °C for 5 min. To the reaction mixture, 200 μl of 2.7 mmol/l S-2222, a synthetic substrate for Factor Xa, was added and this was incubated for a further 5 min. The reaction was stopped by the addition of 100 μl of 50% acetic acid, and the absorbance for liberation of* p-*nitroaniline was read at 405 nm.*

addition of EDTA. We also examined CPB I binding to commercially available phospholipids such as phosphatidyl-ethanolamine, -serine, -inositol, -choline, cardiolipin and sphingomyelin. CPB I bound to each of the acidic phospholipids in the presence of $CaCl_2$, but there was only weak binding to phosphatidylcholine and sphingomyelin.

## CPB I in human organs

The presence of CPB I was demonstrated by Western blotting in EGTA extracts of various human organs including placenta, lung, kidney, liver, spleen, pancreas, adrenal gland and intestine. In a histochemical analysis,

**Fig. 5. Effect of CPB I on factor X activation by extrinsic pathway or Russell's viper venom (RVV).**

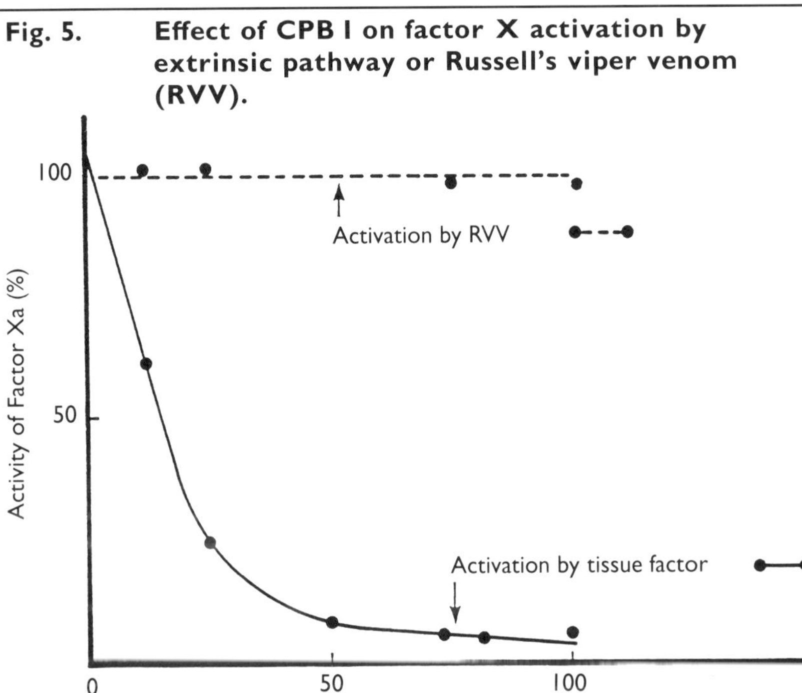


*Factor X (200 μl of 0.75 units/ml) was activated by incubation with a reaction mixture of 10 μl of 120 units/ml Factor VII and 100 μl of tissue thromboplastin-$CaCl_2$ (Lyoplastin), or by incubation with 110 μl of 0.066 mg/ml RVV/ 100 mmol $CaCl_2$ mixture at 37 °C for 3 min in the presence of 100 μl of various concentrations of CPB I. To the reaction mixture, 200 μl of 0.15 μM-S2337, a synthetic substrate for Factor Xa, was added and this was incubated for a further 3 min. The reaction was stopped by the addition of 200 μl of 50% acetic acidl, and the absorbance for* p-*nitroaniline was read at 405 nm.*

the surface of the chorionic villi was stained by the peroxidase–antiperoxidase complex method using an anti-CPB I rabbit polyclonal antibody. Endothelial cells, platelets, macrophages, decidua cells, pancreatic duct, bile duct, adrenal gland and prostate gland also stained positively. Electron microscopy studies showed that CPB I was located on the placental microvilli which contain the antithrombotic factors such as thrombomodulin and platelet-aggregation inhibitor.

The CPB I content in various body fluids determined by e.l.i.s.a. is shown in Table 1. CPB I was present in the plasma at a concentration

**Fig. 6. Effect of CPB I on prothrombin activation by Factor Xa or *Echis carinatus* venom (Ecarin).**

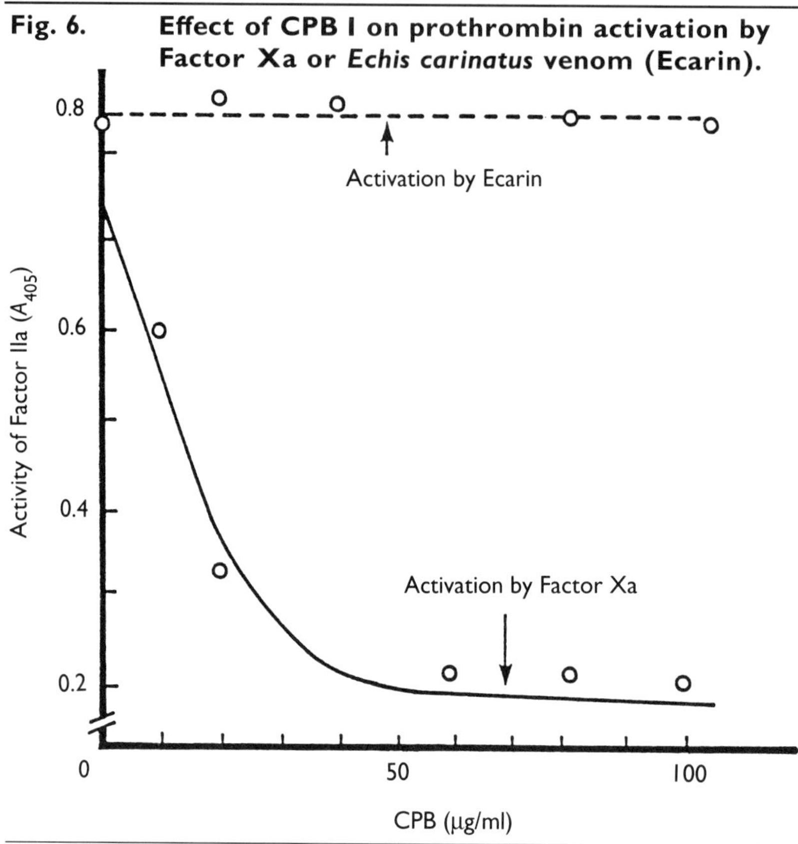


*Prothrombin (50 μl of 0.5 units/ml) was activated by incubation with 100 μl of 0.4 μg/ml Factor Xa, 50 μl of 3ISY1.8 μg/ml phospholipid and 50 μl of 0.1 M-$CaCl_2$, or by incubation with 100 μl of 1 mu/ml Ecarin in the presence or absence of CPB I at 37 °C for 3 min. To each reaction mixture, 50 μl of 1.37 mg/ml RM-60 1-2, a synthetic substrate for thrombin, was added, and this was incubated for a further 15 min. The reaction was stopped by the addition of 550 μl of 50% acetic acid, and the absorbance at 405 nm was read.*

of about 10 ng/ml or less. This level did not change during normal pregnancy, regardless of gestational age, but it increased slightly in patients with toxaemia of pregnancy and disseminated intravascular coagulation, and in the umbilical cord blood. Blood samples obtained at necropsy showed markedly higher levels of CPB I. Normal ascites fluid contained a high level, and this increased in carcinomatous ascites. The level of CPB I was high in amniotic fluid, but remained constant during

**Fig. 7. Effect of CPB I on binding of prothrombin to phospholipid vesicles.**

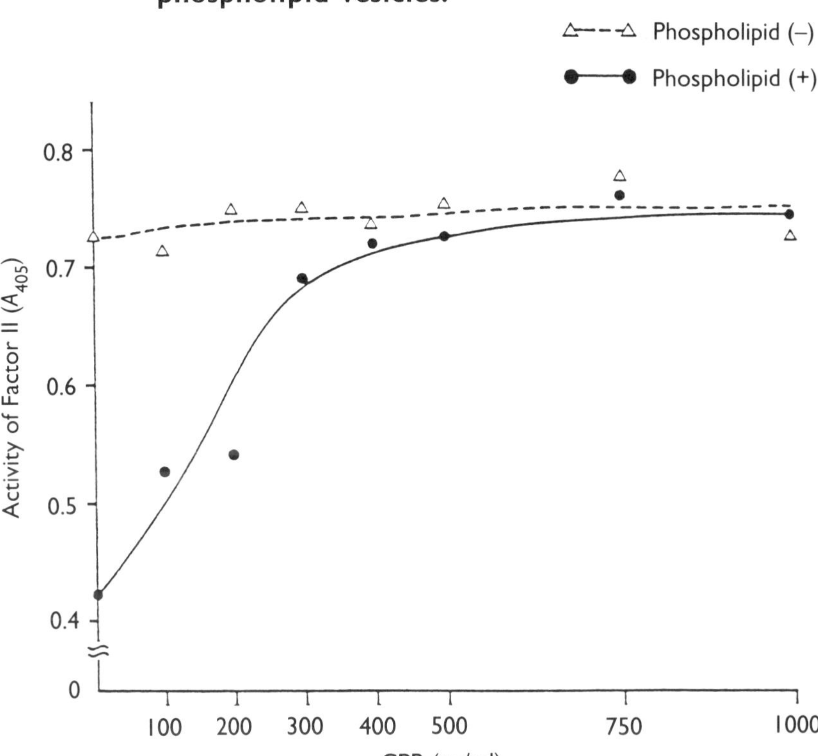


*A reaction mixture of 50 μl of 0.5 units/ml prothrombin, 50 μl of 0.1 M $CaCl_2$, 50 μl of 1% BSA, and 50 μl of various concentrations of CPB I in the presence or absence of 100 μl of lipsome, was incubated at 37 °C for 5 min. The mixture was centrifuged at 100 000 g. Unbound prothrombin in the supernatant was measured by RM 60 1-2 after activation with Ecarin.*

gestation. Human milk and seminal fluid were rich in CPB I. The content of CPB I in ovarian follicular fluid was also high, especially in immature follicles and atretic follicles.

**Table 1. CPB content in body fluid samples**

| Samples | Number | Content (mean ± SD, ng/ml) |
|---|---|---|
| Blood plasma samples | | |
| Nonpregnant | 15 | 10.0 ± 4.9 |
| Pregnant | 17 | 6.8 ± 2.1 |
| Toxemia | 46 | 4.7 ± 7.2 |
| Atonic bleeding | 8 | 11.4 ± 6.0 |
| Umbilical cord | 48 | 40.7 ± 69.0 |
| DIC | 12 | 25.4 ± 52.5 |
| Necropsy | 6 | 969.4 ± 1478.0 |
| Other body fluids | | |
| Normal ascites | 14 | 129.1 ± 50.4 |
| Carcinomatous ascites (ovary) | 25 | 495.4 ± 1060.9 |
| Carcinomatous ascites (GI) | 6 | 713.0 ± 622.2 |
| Carcinomatous hydrothorax | 3 | 24.0 ± 16.6 |
| Cerebro-spinal fluid | 10 | 10.0 ± 14.8 |
| Follicular fluid | | |
| Mature | 33 | 29 ± 37 |
| Immature | 18 | 72 ± 74 |
| Atretic | 14 | 102 ± 80 |
| Milk | 3 | 262.6 ± 131.1 |
| Seminal fluid | 87 | 9010 ± 4010 |

## Discussion

Many proteins with $Ca^{2+}$-phospholipid-binding activity have been discovered in various species including plants, and have been reported using different names. We reported two anti-thrombotic factors, a coagulation inhibitor and a platelet-aggregation inhibitor, isolated from the human placenta [5]. After amino acid sequencing of the placental coagulation inhibitor (CPB I), we noticed that this protein was a member of the annexin family with the characteristic $Ca^{2+}$-phospholipid-binding capacity. In an effort to alleviate growing confusion in this field, the generic term annexin was adopted with the lipocortin numbering [14].

A basic property of annexins is $Ca^{2+}$-dependent phospholipid-binding, raising the possibility of various biological roles such as inhibition of phospholipases $A_2$ and C, inhibition of $Ca^{2+}$-phospholipid-dependent process of blood coagulation, binding to membranes, actin or cytoskeleton, aggregation of chromaffin granules, exocytosis, regulation of $Ca^{2+}$ mobilization and cell signalling and differentiation. Annexin III was identified as an enzyme of inositol phosphate metabolism. Among these various activities, it is known which function(s) is/are biologically essential.

In considering the inhibitory action of annexins on blood coagulation, it must be taken into account that annexin family proteins exert both anticoagulant and antiphospholipase $A_2$ activities by interacting with phospholipids in the presence of $Ca^{2+}$ [6,15]. The inhibitory activity of annexins on phospholipase $A_2$ is almost completely abolished by supplementation of phospholipid as substrate. The inhibition of blood coagulation by CPB I was decreased by increasing the amount of phospholipid, but did not disappear completely as in phospholipase $A_2$. This may be caused by inhibition of binding of Factor X or prothrombin to phospholipid by CPB I, in addition to the direct activity of $Ca^{2+}$-phospholipid-binding of CPB I.

Endothelial cells contain materials that cross-react with anti-CPB I and CPB II antibodies. These are normally intracellular components, and not cell-surface membrane-bound proteins like thrombomodulin. Therefore, they seem unlikely to be the direct anti-thrombotic factors in the blood vessels. However, they may play an anti-thrombotic role for the following reasons. When the umbilical artery and veins were incised, cut and opened, and immersed in saline, CPB I appeared gradually in the saline and reached maximum concentration after 2–3 h. This indicates that endothelial cells, either loosing their viability or dying, release CPB I from the intracellular to the extracellular space. High levels of CPB I in necropsy blood and in various cavity fluids can thus be explained by the release of CPB I from dead endothelial and exfoliated cavity cells. At the same time, it is known that tissue thromboplastin is released from dying or dead cells. It is possible, therefore, to suggest that CPB I, and probably the other annexins, might play a defensive role in anti-thrombotic mechanisms by neutralizing tissue thromboplastin activity released from damaged cells. On the other hand, annexins (at least CPB I) located on the chorionic villi may have an anti-thrombotic function together with thrombomodulin and the other anti-thrombotic mechanisms.

*This work was supported by a Grant from Committee of the Science Council of the Japan Education Department.*

## References

1. Myatt, L. & Elder, M. G. (1977) Nature (London) **268**, 159–160
2. Hutton, R. A., Dandona, P., Chow, F. P. R. & Craft, I. L. (1980) Thromb. Res. **17**, 465–471
3. Maruyama, I., Bell, C. E. & Majerus, P. W. (1985) J. Cell Biol. **101**, 363–371
4. Kobayashi, T. & Terao, T. (1985) Acta Obstet. Gynecol. Jpn. (Engl. Ed.) **37**, 783–792
5. Maki, M., Murata, M. & Shidara, Y. (1984) Eur. J. Obstet. Gynecol. Reprod. Biol. **17**, 149–154

6. Iwasaki, A., Suda, M., Nakao, H., Nagoya, T., Saino, Y., Arai, K., Mizoguchi, T., Sato, F., Yoshizaki, H., Hirata, M., Miyata, T., Shidara, Y., Murata, M. & Maki, M. (1987) J. Biochem. (Tokyo) **102**, 1261–1273
7. Funakowshi, T., Hendrickson, L. E., McMullen, B. A. & Fujikawa, K. (1987) Biochemistry **26**, 8087–8092
8. Grundmann, U., Abel, K-J., Bohn, H., Löbermann, H., Lottspeich, F. & Kupper, H. (1988) Proc. Natl. Acad. Sci. U.S.A. **85**, 3708–3712
9. Maki, M., Shidara, Y., Murata, M., Takahashi, O., Narita, M., Kume, K., Saino, Y., Nagoya, T., Yoshizaki, H., Iwasaki, A. & Nakao, H. (1988) J. Obst. Gynecol. Neonat. Hematol. (Jpn) **12**, 41–48
10. Goto, K. & Sato, H. (1980) Tohoku J. Exp. Med. **131**, 399–407
11. Iwasaki, A., Suda, M., Watanabe, M., Nakao, H., Hattori, Y., Nagoya, T., Saino, Y., Shidara, Y. & Maki, M. (1989) J. Biochem. (Tokyo) **106**, 43–49
12. Yoshizaki, H., Mizoguchi, T., Arai, K., Shiratsuchi, M., Shidara, Y. & Maki, M. (1990) J. Biochem. (Tokyo) **107**, 43–50
13. Maki, M. (1991) in Haemostasis and Thrombosis in Obstetrics and Gynecology, pp. 105–123, Die Medizimische Verlangsgesellschaft mbH, Marburg
14. Crumpton, M. J. & Dedman, J. R. (1990) Nature (London) **345**, 212
15. Chap, H., Comfurius, P., Bevers, E. M., Fauvel, J., Vicendo, P., Blazy, L. D. & Zwaal, R. F. A. (1988) Biochem. Biophys. Res. Commun. **150**, 972–978

# Index